# Air Pollution

# Air Pollution

Hunter Dawson

Larsen & Keller
www.larsen-keller.com

Air Pollution
Hunter Dawson
ISBN: 978-1-64172-670-2 (Hardback)

# ⊟ Larsen & Keller

Published by Larsen and Keller Education,
5 Penn Plaza,
19th Floor,
New York, NY 10001, USA

**Cataloging-in-Publication Data**

Air pollution / Hunter Dawson.
        p. cm.
Includes bibliographical references and index.
ISBN 978-1-64172-670-2
1. Air--Pollution. 2. Pollution. 3. Environmental protection. I. Dawson, Hunter.
TD883 .A47 2022
363.739 2--dc23

For more information regarding Larsen and Keller Education and its products, please visit the publisher's website www.larsen-keller.com

# Table of Contents

# Preface

This book has been written, keeping in view that students want more practical information. Thus, my aim has been to make it as comprehensive as possible for the readers. I would like to extend my thanks to my family and co-workers for their knowledge, support and encouragement all along.

Air pollution is a condition when the Earth's atmosphere or environment contains excessive quantities of harmful substances including gases, particles and biological molecules. The sources of air pollution can be broadly divided into natural sources and man-made sources. Some of the natural sources are dust, methane emitted by animals and radon gas from radioactive decay in the Earth's crust. A few of the man-made sources are fossil fuel power stations and motor vehicles. Air pollution can cause serious diseases, allergies and even death among humans. It also harms other living organisms like animals and affects food crops. Air pollution can also damage natural and built environment. The topics included in this textbook on air pollution are of utmost significance and bound to provide incredible insights to readers. Different approaches, evaluations and studies related to this problem have been included in this book. It will serve as a valuable source of reference for those interested in this field.

A brief description of the chapters is provided below for further understanding:

Chapter – Introduction

Air pollution refers to the presence of harmful or excessive quantities of substances like gases, particles and biological molecules into the Earth's atmosphere. Some of the areas of study related to air pollution are air pollution forecasting and air pollution hotspots. This is an introductory chapter which will introduce briefly all the significant aspects of air pollution.

Chapter – Air Pollutants

Some of the different types of air pollutants are toxic air pollutants, criteria air pollutants and indoor air pollutants. A few examples of air pollutants are sulfur dioxide, nitrogen oxide, ozone and carbon monoxide. This chapter has been carefully written to provide an easy understanding of these types of air pollutants and air pollution dispersion model.

Chapter – Effects of Air Pollution

Air pollution adversely affects the health in humans, especially harming the lungs and brain. Some of its other effects include arctic haze, global warming, ozone depletion and acid rain. The topics elaborated in this chapter will help in gaining a better perspective about these effects of air pollution.

Chapter – Air Quality Monitoring, Measurement and Health Index

Air quality monitoring refers to the measurement of air quality using various techniques such as air sampling. Air quality health index is a scale which is used to analyze the impact which air quality has on health. The diverse aspects of air quality monitoring as well as air quality health index have been thoroughly discussed in this chapter.

Chapter – Control and Prevention

Air pollution is controlled using various devices that prevent harmful gaseous and solid pollutants from entering the atmosphere. A few examples of such devices are dust collectors, scrubbers and thermal oxidizers. This chapter has been carefully written to provide an easy understanding of these types of air pollution control devices.

**Hunter Dawson**

# 1
# Introduction

Air pollution refers to the presence of harmful or excessive quantities of substances like gases, particles and biological molecules into the Earth's atmosphere. Some of the areas of study related to air pollution are air pollution forecasting and air pollution hotspots. This is an introductory chapter which will introduce briefly all the significant aspects of air pollution.

## Air Pollution

Air pollution is the human introduction into the atmosphere of chemicals, particulate matter, or biological materials that cause harm or discomfort to humans or other living organisms, or damages the environment. Air pollution causes deaths and respiratory disease. Air pollution is often identified with major stationary sources, but the greatest source of emissions is mobile sources, mainly automobiles. Gases such as carbon dioxide, which contribute to global warming, have recently been labeled as pollutants by climate scientists, while they also recognize that carbon dioxide is essential for plant life through photosynthesis.

The atmosphere is a complex, dynamic natural gaseous system that is essential to support life on planet Earth. Stratospheric ozone depletion due to air pollution has long been recognized as a threat to human health as well as to the Earth's ecosystems.

### Pollutants

There are many substances in the air which may impair the health of plants and animals (including humans), or reduce visibility. These arise both from natural processes and human activity.

Substances not naturally found in the air or at greater concentrations or in different locations from usual are referred to as pollutants.

Pollutants can be classified as either primary or secondary. Primary pollutants are substances directly emitted from a process, such as ash from a volcanic eruption, the carbon monoxide gas from a motor vehicle exhaust or sulfur dioxide released from factories.

Secondary pollutants are not emitted directly. Rather, they form in the air when primary pollutants react or interact. An important example of a secondary pollutant is ground level ozone - one of the many secondary pollutants that make up photochemical smog.

Note that some pollutants may be both primary and secondary: that is, they are both emitted directly and formed from other primary pollutants.

Major primary pollutants produced by human activity include:

- Sulfur oxides ($SO_x$) especially sulfur dioxide are emitted from burning of coal and oil.

- Nitrogen oxides ($NO_x$) especially nitrogen dioxide are emitted from high temperature combustion. Can be seen as the brown haze dome above or plume downwind of cities.

- Carbon monoxide is colorless, odorless, non-irritating but very poisonous gas. It is a product by incomplete combustion of fuel such as natural gas, coal or wood. Vehicular exhaust is a major source of carbon monoxide.

- Carbon dioxide ($CO_2$), a greenhouse gas emitted from combustion.

- Volatile organic compounds (VOC), such as hydrocarbon fuel vapors and solvents.

- Particulate matter (PM), measured as smoke and dust. $PM_{10}$ is the fraction of suspended particles 10 micrometers in diameter and smaller that will enter the nasal cavity. $PM_{2.5}$ has a maximum particle size of 2.5 µm and will enter the bronchi's and lungs.

- Toxic metals, such as lead, cadmium and copper.

- Chlorofluorocarbons (CFCs), harmful to the ozone layer emitted from products currently banned from use.

- Ammonia ($NH_3$) emitted from agricultural processes.

- Odors, such as from garbage, sewage, and industrial processes.

- Radioactive pollutants produced by nuclear explosions and war explosives, and natural processes such as radon.

Secondary pollutants include:

- Particulate matter formed from gaseous primary pollutants and compounds in photochemical smog, such as nitrogen dioxide.

- Ground level ozone ($O_3$) formed from $NO_x$ and VOCs.

- Peroxyacetyl nitrate (PAN) similarly formed from $NO_x$ and VOCs.

Minor air pollutants include:

- A large number of minor hazardous air pollutants. Some of these are regulated in USA under the Clean Air Act and in Europe under the Air Framework Directive.

- A variety of persistent organic pollutants, which can attach to particulate matter.

## Sources of Pollutants

Dust storm approaching Stratford, Texas.

Using a controlled burn on a field in South Georgia in preparation for spring planting.

Puxi area of Shanghai at sunset. The sun has not actually dropped below the horizon yet, rather it has reached the smog line.

Sources of air pollution refer to the various locations, activities or factors which are responsible for the releasing of pollutants in the atmosphere. These sources can be classified into two major categories which are:

- Anthropogenic sources (human activity) mostly related to burning different kinds of fuel.

- "Stationary Sources" as smoke stacks of power plants, manufacturing facilities, municipal waste incinerators.

- "Mobile Sources" as motor vehicles, aircraft etc.

- Marine vessels, such as container ships or cruise ships, and related port air pollution.

- Burning wood, fireplaces, stoves, furnaces and incinerators.

- Oil refining, and industrial activity in general.

- Chemicals, dust and controlled burn practices in agriculture and forestry management.

- Fumes from paint, hair spray, varnish, aerosol sprays and other solvents.

- Waste deposition in landfills, which generate methane.

- Military, such as nuclear weapons, toxic gases, germ warfare and rocketry.

## Natural Sources

- Dust from natural sources, usually large areas of land with little or no vegetation.

- Methane, emitted by the digestion of food by animals, for example cattle.

- Radon gas from radioactive decay within the Earth's crust.

- Smoke and carbon monoxide from wildfires.

- Volcanic activity, which produce sulfur, chlorine, and ash particulates.

## Emission Factors

Air pollutant emission factors are representative values that attempt to relate the quantity of a pollutant released to the ambient air with an activity associated with the release of that pollutant. These factors are usually expressed as the weight of pollutant divided by a unit weight, volume, distance, or duration of the activity emitting the pollutant (e.g., kilograms of particulate emitted per megagram of coal burned). Such factors facilitate estimation of emissions from various sources of air pollution. In most cases, these factors are simply averages of all available data of acceptable quality, and are generally assumed to be representative of long-term averages.

The United States Environmental Protection Agency has published a compilation of air pollutant emission factors for a multitude of industrial sources. Other countries have published similar compilations, as has the European Environment Agency.

## Indoor Air Quality

A lack of ventilation indoors concentrates air pollution where people often spend the majority of their time. Radon (Rn) gas, a carcinogen, is exuded from the Earth in certain locations and trapped inside houses. Building materials including carpeting and plywood emit formaldehyde ($H_2CO$) gas. Paint and solvents give off volatile organic compounds (VOCs) as they dry. Lead paint can degenerate into dust and be inhaled. Intentional air pollution is introduced with the use of air fresheners, incense, and other scented items. Controlled wood fires in stoves and fireplaces can add significant amounts of smoke particulates into the air, inside and out. Indoor pollution fatalities may be caused by using pesticides and other chemical sprays indoors without proper ventilation.

Carbon monoxide (CO) poisoning and fatalities are often caused by faulty vents and chimneys, or by the burning of charcoal indoors. Chronic carbon monoxide poisoning can result even from poorly adjusted pilot lights. Traps are built into all domestic plumbing to keep sewer gas, hydrogen sulfide, out of interiors. Clothing emits tetrachloroethylene, or other dry cleaning fluids, for days after dry cleaning.

Though its use has now been banned in many countries, the extensive use of asbestos in industrial and domestic environments in the past has left a potentially very dangerous material in many localities. Asbestosis is a chronic inflammatory medical condition affecting the tissue of the lungs. It occurs after long-term, heavy exposure to asbestos from asbestos-containing materials in structures. Sufferers have severe dyspnea (shortness of breath) and are at an increased risk regarding several different types of lung cancer. As clear explanations are not always stressed in non-technical literature, care should be taken to distinguish between several forms of relevant diseases. According to the World Health Organization (WHO), these may defined as; asbestosis, lung cancer, and mesothelioma (generally a very rare form of cancer, when more widespread it is almost always associated with prolonged exposure to asbestos).

Biological sources of air pollution are also found indoors, as gases and airborne particulates. Pets produce dander, people produce dust from minute skin flakes and decomposed hair, dust mites in bedding, carpeting and furniture produce enzymes and micrometer-sized fecal droppings, inhabitants emit methane, mold forms in walls and generates mycotoxins and spores, air conditioning systems can incubate Legionnaires' disease and mold, and houseplants, soil and surrounding gardens can produce pollen, dust, and mold. Indoors, the lack of air circulation allows these airborne pollutants to accumulate more than they would otherwise occur in nature.

## Health Effects

Health effects caused by air pollutants may range from subtle biochemical and physiological changes to difficulty in breathing, wheezing, coughing, and aggravation of existing respiratory and cardiac conditions. These effects can result in increased medication use, increased doctor or emergency room visits, more hospital admissions and premature death. The human health effects of poor air quality are far-reaching, but principally affect the body's respiratory system and the cardiovascular system. Individual reactions to air pollutants depend on the type of pollutant a person is exposed to, the degree of exposure, the individual's health status, and genetics.

The World Health Organization states that 2.4 million people die each year from causes directly attributable to air pollution, with 1.5 million of these deaths attributable to indoor air pollution. A study by the University of Birmingham has shown a strong correlation between pneumonia-related deaths and air pollution from motor vehicles. Direct causes of deaths related to air pollution include aggravated asthma, bronchitis, emphysema, lung and heart diseases, and respiratory allergies.

The worst short-term civilian pollution crisis in India was the 1984 Bhopal Disaster. Leaked industrial vapors from the Union Carbide factory, belonging to Union Carbide, Inc., U.S.A., killed more than 2,000 people outright and injured anywhere from 150,000 to 600,000 others, some 6,000 of whom would later die from their injuries. The United Kingdom suffered its worst air pollution event when the December 4 Great Smog of 1952 formed over London. An accidental leak of anthrax spores from a biological warfare laboratory in the former USSR in 1979 near Sverdlovsk is believed to have been the cause of hundreds of civilian deaths. The worst single incident of air pollution to occur in the United States of America occurred in Donora, Pennsylvania in late October, 1948, when 20 people died and over 7,000 were injured.

## Effects on Children

In cities around the world with high levels air pollutants, children have a higher probability of developing asthma, pneumonia, and other lower respiratory infections. Because children spend more time outdoors and have higher minute ventilation, they are more susceptible to the dangers of air pollution.

Research by the World Health Organization shows that the highest concentrations of particulate matter can be found in countries with low economic strength and high poverty and population rates. Examples of these countries include Egypt, Sudan, Mongolia, and Indonesia. Protective measures to ensure the health of youth are being undertaken in cities such as New Delhi, where buses now use compressed natural gas to help eliminate the "pea-soup" fog.

In the U.S., the Clean Air Act was passed in 1970. However, in 2002, at least 146 million Americans were living in areas that did not meet at least one of the "criteria pollutants" laid out in the 1997 National Ambient Air Quality Standards. Those pollutants included: ozone, particulate matter, sulfur dioxide, nitrogen dioxide, carbon monoxide, and lead.

## Cystic Fibrosis

Cystic fibrosis patients are born with decreased lung function. For them, everyday pollutants such as smoke emissions from automobiles, tobacco smoke, and improper use of indoor heating devices can more severely affect lung function.

A study from 1999 to 2000 by the University of Washington showed that patients near and around particulate matter air pollution had an increased risk of pulmonary exacerbations and decrease in lung function. Patients were examined before the study for amounts of specific pollutants like P. aeruginosa or B. cepacia, as well as their socioeconomic standing. During the time of the study, 117 deaths were associated with air pollution. A trend was noticed that patients living in large metropolitan areas had higher level of pollutants in their system because of greater emission levels in larger cities.

## Chronic Obstructive Pulmonary Disease

Chronic obstructive pulmonary disease (COPD) includes illnesses such as chronic bronchitis, emphysema, and some forms of asthma. Two researchers, Holland and Reid, conducted research on 293 male postal workers in London during the time of the Great Smog of 1952 and 477 male postal workers in the rural setting. The amount of the pollutant FEV1 was significantly lower in urban employees however lung function was decreased due to city pollutions such as car fumes and increased amount of cigarette exposure.

It is believed that, much like cystic fibrosis, serious health problems become more apparent among people living in a more urban environment. Studies have shown that in urban areas, patients suffer mucus hypersecretion, lower levels of lung function, and more self-diagnosis of chronic bronchitis and emphysema.

## Great Smog of 1952

In a span of four days, a combination of dense fog and sooty black coal smoke covered the London area. The fog was so dense that residents of London could not see in front of them. The extreme reduction in visibility was accompanied by an increase in criminal activity as well as transportation delays and a virtual shutdown of the city. During the four-day period of the fog, 12,000 people are believed to have been killed.

## Environmental Impacts

The greenhouse effect is a phenomenon whereby greenhouse gases create a condition in the upper atmosphere causing a trapping of heat and leading to increased surface and lower tropospheric temperatures. It shares this property with many other gases, the largest overall forcing on Earth coming from water vapor. Other greenhouse gases include methane, hydrofluorocarbons, perfluorocarbons, chlorofluorocarbons, $NO_x$, and ozone. Many greenhouse gases, contain carbon, and some of that from fossil fuels.

This effect has been understood by scientists for about a century, and technological advancements during this period have helped increase the breadth and depth of data relating to the phenomenon. Currently, scientists are studying the role of changes in composition of greenhouse gases from natural and anthropogenic sources for the effect on climate change.

A number of studies have also investigated the potential for long-term rising levels of atmospheric carbon dioxide to cause slight increases in the acidity of ocean waters and the possible effects of this on marine ecosystems. However, carbonic acid is a very weak acid, and is utilized by marine organisms during photosynthesis.

## Reduction Efforts

There are various air pollution control technologies and urban planning strategies available to reduce air pollution.

Efforts to reduce pollution from mobile sources includes primary regulation (many developing countries have permissive regulations), expanding regulation to new sources (such as cruise and transport ships, farm equipment, and small gas-powered equipment such as lawn trimmers, chainsaws, and snowmobiles), increased fuel efficiency (such as through the use of hybrid vehicles), conversion to cleaner fuels (such as bioethanol, biodiesel, or conversion to electric vehicles).

## Control Devices

The following items are commonly used as pollution control devices by industry or transportation devices. They can either destroy contaminants or remove them from an exhaust stream before it is emitted into the atmosphere.

- Particulate control:
  - Mechanical collectors (dust cyclones, multicyclones).
  - Electrostatic precipitators.

- Baghouses.
- Particulate scrubbers.
- Scrubbers:
  - Baffle spray scrubber.
  - Cyclonic spray scrubber.
  - Ejector venturi scrubber.
  - Mechanically aided scrubber.
  - Spray tower.
  - Wet scrubber.
- $NO_x$ control:
  - Low $NO_x$ burners.
  - Selective catalytic reduction (SCR).
  - Selective non-catalytic reduction (SNCR).
  - NOx scrubbers.
  - Exhaust gas recirculation.
  - Catalytic converter (also for VOC control).
- VOC abatement:
  - Adsorption systems, such as activated carbon.
  - Flares.
  - Thermal oxidizers.
  - Catalytic oxidizers.
  - Biofilters.
  - Absorption (scrubbing).
  - Cryogenic condensers.
  - Vapor recovery systems.
- Acid Gas/$SO_2$ control:
  - Wet scrubbers.
  - Dry scrubbers.

- ○   Flue gas desulfurization.

- •   Mercury control:

    - ○   Sorbent Injection Technology.

    - ○   Electro-Catalytic Oxidation (ECO).

    - ○   K-Fuel.

- •   Dioxin and furan control.

- •   Miscellaneous associated equipment:

    - ○   Source capturing systems.

    - ○   Continuous emissions monitoring systems (CEMS).

## Atmospheric Dispersion Models

The basic technology for analyzing air pollution is through the use of a variety of mathematical models for predicting the transport of air pollutants in the lower atmosphere. The principal methodologies are noted below:

- •   Point source dispersion, used for industrial sources.

- •   Line source dispersion, used for airport and roadway air dispersion modeling.

- •   Area source dispersion, used for forest fires or duststorms.

- •   Photochemical models, used to analyze reactive pollutants that form smog.

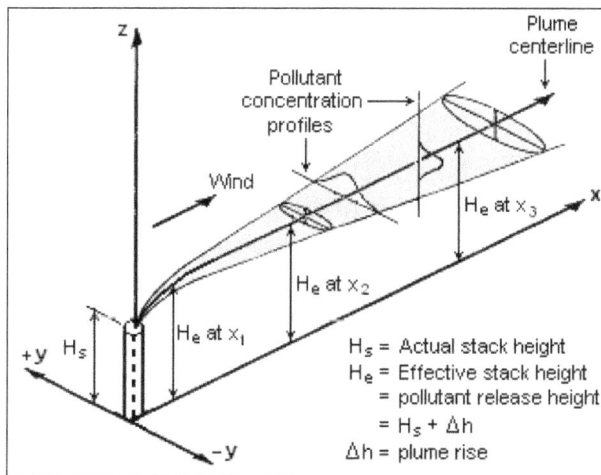

Visualization of a buoyant Gaussian air pollution dispersion plume
as used in many atmospheric dispersion models.

The point source problem is the best understood, since it involves simpler mathematics and has been studied for a long period of time, dating back to about the year 1900. It uses a Gaussian dispersion model to forecast air pollution plumes, with consideration given to wind velocity, stack height, emission rate and stability class (a measure of atmospheric turbulence).

The roadway air dispersion model was developed starting in the late 1950s and early 1960s in response to requirements of the National Environmental Policy Act and the U.S. Department of Transportation (then known as the Federal Highway Administration) to understand impacts of proposed new highways upon air quality, especially in urban areas. Several research groups were active in this model development, among which were the Environmental Research and Technology (ERT) group in Lexington, Massachusetts, the ESL Inc. group in Sunnyvale, California and the California Air Resources Board group in Sacramento, California.

Area source models were developed in 1971 through 1974 by the ERT and ESL groups, but addressed a smaller fraction of total air pollution emissions, so that their use and need was not as widespread as the line source model, which enjoyed hundreds of different applications as early as the 1970s.

Likewise, photochemical models were developed primarily in the 1960s and 1970s. Their use was constrained to regional needs, such as understanding smog formation in Los Angeles, California.

## Legal Regulations in some Nations

Smog in Cairo

In general, there are two types of air quality standards. The first class of standards (such as the U.S. National Ambient Air Quality Standards) set maximum atmospheric concentrations for specific pollutants. Environmental agencies enact regulations which are intended to result in attainment of these target levels. The second class (such as the North American Air Quality Index) take the form of a scale with various thresholds, which is used to communicate to the public the relative risk of outdoor activity. The scale may or may not distinguish between different pollutants.

## Canada

In Canada, air quality is typically evaluated against standards set by the Canadian Council of Minister for the Environment (CCME), an inter-governmental body of federal, provincial and territorial Ministers responsible for the environment. The CCME set Canada Wide Standards(CWS).

## European Union

National Emission Ceilings (NEC) for certain atmospheric pollutants are regulated by Directive 2001/81/EC (NECD). As part of the preparatory work associated with the revision of the NECD, the European Commission is assisted by the NECPI working group (National Emission Ceilings – Policy Instruments).

## United Kingdom

Air quality targets set by the UK's Department for Environment, Food and Rural Affairs (DEFRA) are mostly aimed at local government representatives responsible for the management of air quality in cities, where air quality management is the most urgent. The UK has established an air quality network where levels of the key air pollutants are published by monitoring centers. Air quality in Oxford, Bath and London is particularly poor. One controversial study performed by the Calor Gas company and published in the Guardian newspaper compared walking in Oxford on an average day to smoking over sixty light cigarettes.

More precise comparisons can be collected from the UK Air Quality Archive which allows the user to compare a cities management of pollutants against the national air quality objectives set by DEFRA in 2000.

Localized peak values are often cited, but average values are also important to human health. The UK National Air Quality Information Archive offers almost real-time monitoring of "current maximum" air pollution measurements for many UK towns and cities. This source offers a wide range of constantly updated data, including:

- Hourly Mean Ozone ($\mu g/m^3$).

- Hourly Mean Nitrogen dioxide ($\mu g/m^3$).

- Maximum 15-Minute Mean Sulphur dioxide ($\mu g/m^3$).

- 8-Hour Mean Carbon monoxide ($mg/m^3$).

- 24-Hour Mean PM10 ($\mu g/m^3$ Grav Equiv).

DEFRA acknowledges that air pollution has a significant effect on health and has produced a simple banding index system is used to create a daily warning system that is issued by the BBC Weather Service to indicate air pollution levels. DEFRA has published guidelines for people suffering from respiratory and heart diseases.

## United States

Looking down from the Hollywood Hills, with Griffith Observatory on the hill in the foreground, air pollution is visible in downtown Los Angeles on a late afternoon.

In the 1960s, 1970s, and 1990s, the United States Congress enacted a series of Clean Air Acts which significantly strengthened regulation of air pollution. Individual U.S. states, some European nations and eventually the European Union followed these initiatives. The Clean Air Act sets numerical limits on the concentrations of a basic group of air pollutants and provide reporting and enforcement mechanisms.

In 1999, the United States EPA replaced the Pollution Standards Index (PSI) with the Air Quality Index (AQI) to incorporate new PM2.5 and Ozone standards.

The effects of these laws have been very positive. In the United States between 1970 and 2006, citizens enjoyed the following reductions in annual pollution emissions:

- Carbon monoxide emissions fell from 197 million tons to 89 million tons.
- Nitrogen oxide emissions fell from 27 million tons to 19 million tons.
- Sulfur dioxide emissions fell from 31 million tons to 15 million tons.
- Particulate emissions fell by 80 percent.
- Lead emissions fell by more than 98 percent.

The EPA proposed, in June 2007, a new threshold of 75 ppb. This falls short of the scientific recommendation, but is an improvement over the current standard.

Polluting industries are lobbying to keep the current (weaker) standards in place. Environmentalists and public health advocates are mobilizing to support compliance with the scientific recommendations.

The National Ambient Air Quality Standards are pollution thresholds which trigger mandatory remediation plans by state and local governments, subject to enforcement by the EPA.

## Indoor Air Pollution

It refers to the physical, chemical, and biological characteristics of air in the indoor environment within a home, building, or an institution or commercial facility. Indoor air pollution is a concern in the developed countries, where energy efficiency improvements sometimes make houses relatively airtight, reducing ventilation and raising pollutant levels. Indoor air problems can be subtle and do not always produce easily recognized impacts on health. Different conditions are responsible for indoor air pollution in the rural areas and the urban areas.

In the developing countries, it is the rural areas that face the greatest threat from indoor pollution, where some 3.5 billion people continue to rely on traditional fuels such as firewood, charcoal, and cowdung for cooking and heating. Concentrations of indoor pollutants in households that burn traditional fuels are alarming. Burning such fuels produces large amount of smoke and other air pollutants in the confined space of the home, resulting in high exposure. Women and children are the groups most vulnerable as they spend more time indoors and are exposed to the smoke. In 1992, the World Bank designated indoor air pollution in the developing countries as one of the four most critical global environmental problems. Daily averages of pollutant level emitted indoors often exceed current WHO guidelines and acceptable levels. Although many hundreds of separate chemical agents have been identified in the smoke from biofuels, the four most serious pollutants are particulates, carbon monoxide, polycyclic organic matter, and formaldehyde. Unfortunately, little monitoring has been done in rural and poor urban indoor environments in a manner that is statistically rigorous.

In urban areas, exposure to indoor air pollution has increased due to a variety of reasons, including the construction of more tightly sealed buildings, reduced ventilation, the use of synthetic materials for building and furnishing and the use of chemical products, pesticides, and household care products. Indoor air pollution can begin within the building or be drawn in from outdoors. Other than nitrogen dioxide, carbon monoxide, and lead, there are a number of other pollutants that affect the air quality in an enclosed space.

Volatile organic compounds originate mainly from solvents and chemicals. The main indoor sources are perfumes, hair sprays, furniture polish, glues, air fresheners, moth repellents, wood preservatives, and many other products used in the house. The main health effect is the imitation of the eye, nose and throat. In more severe cases there may be headaches, nausea and loss of coordination. In the long term, some of the pollutants are suspected to damage to the liver and other parts of the body.

Tobacco smoke generates a wide range of harmful chemicals and is known to cause cancer. It is well known that passive smoking causes a wide range of problems to the passive smoker (the person who is in the same room with a smoker and is not himself/herself a smoker) ranging from burning eyes, nose, and throat irritation to cancer, bronchitis, severe asthma, and a decrease in lung function.

Pesticides, if used carefully and the manufacturers, instructions followed carefully they do not cause too much harm to the indoor air.

Biological pollutants include pollen from plants, mite, hair from pets, fungi, parasites, and some bacteria. Most of them are allergens and can cause asthma, hay fever, and other allergic diseases.

Formaldehyde is a gas that comes mainly from carpets, particle boards, and insulation foam. It causes irritation to the eyes and nose and may cause allergies in some people.

Asbestos is mainly a concern because it is suspected to cause cancer.

Radon is a gas that is emitted naturally by the soil. Due to modern houses having poor ventilation, it is confined inside the house causing harm to the dwellers.

## Smog

Smog is also used to describe the type of fog which has smoke or soot in it. Smog is a yellowish or blackish fog formed mainly by a mixture of pollutants in the atmosphere which consists of fine particles and ground-level ozone. Smog which occurs mainly because of air pollution can also be defined as a mixture of various gases with dust and water vapor. Smog also refers to hazy air that makes breathing difficult.

## How Smog is Formed?

The atmospheric pollutants or gases that form smog are released in the air when fuels are burnt. When sunlight and its heat react with these gases and fine particles in the atmosphere, smog is formed. It is purely caused by air pollution. Ground level ozone and fine particles are released in the air due to complex photochemical reactions between volatile organic compounds (VOC), sulfur dioxide ($SO_2$) and nitrogen oxides ($NO_x$).

These VOC, $SO_2$, and $NO_x$ are called precursors. The main sources of these precursors are pollutants released directly into the air by gasoline and diesel-run vehicles, industrial plants and activities, and heating due to human activities.

Smog is often caused by heavy traffic, high temperatures, sunshine, and calm winds. These are a few of the factors behind an increasing level of air pollution in the atmosphere. During the winter months when the wind speeds are low, it helps the smoke and fog to become stagnate at a place forming smog and increasing pollution levels near the ground closer to where people are respiring. It hampers visibility and disturbs the environment.

The time that smog takes to form depends directly on the temperature. Temperature inversions are situations when warm air does not rise instead stays near the ground. During situations of temperature inversions, if the wind is calm, smog may get trapped and remain over a place for days.

But it is also true that smog is more severe when it occurs farther away from the sources of release of pollutants. This is because the photochemical reactions that cause smog to take place in the air when the released pollutants from heavy traffic drift due to the wind. Smog can thus affect and prove to be dangerous for suburbs, rural areas as well as urban areas or large cities.

## Devastating effects of Smog

Smog is harmful and it is evident from the components that form it and effects that can happen from it. It is harmful to humans, animals, plants, and nature as a whole. Many people deaths were recorded, notably, those relating to bronchial diseases. Heavy smog is responsible for decreasing UV radiation greatly. Thus heavy smog results in low production of the crucial natural element vitamin D leading to cases of rickets among people.

When a city or town gets covered in smog, the effects are felt immediately. Smog can be responsible for any ailment from minor pains to deadly pulmonary diseases such as lung cancer. Smog is well known for causing irritation in the eye. It may also result in inflammation in the tissues of lungs; giving rise to pain in the chest. Other issues or illnesses such as cold and pneumonia are also related to smog. The human body faces great difficulty in defending itself against the harmful effects of smog.

Minor exposure to smog can lead to greater threats of asthma attacks; people suffering from asthma problems must avoid exposure. Smog also causes premature deaths and affects densely populated areas building it up to dangerous levels. The highly affected people include old people, kids and those with cardiac and respiratory complications as they have an easy tendency to be a disadvantage of asthma.

The ground level ozone present in the smog also inhibits plant growth and causes immense damage to crops and forests. Crops, vegetables like soybeans, wheat, tomatoes, peanuts, and cotton are subject to infection when they are exposed to smog. The smog results in mortifying impacts on the environment by killing innumerable animal species and green life as these take time to adapt to breathing and surviving in such toxic environments.

Smog is a devastating problem especially due to the fast modernization or industrialization as the hazardous chemicals involved in smog formation are highly reactive is spread around in the atmosphere. Smoke and sulfur dioxide pollution in urban areas is at much lower levels than in the past, as a result of the law passed to control emissions and in favor of cleaner emission technology.

So how should you fight with the forceful impact of smog? It can be reduced by implementing modifications in your lifestyle, decreasing the consumption of fuels that are non-renewable and by replacing them with alternate sources of fuel which will reduce toxic emissions from vehicles.

## Air Pollution Forecasting

Air pollution forecasting is the application of science and technology to predict the composition of the Air pollution in the atmosphere for a given location and time.

The forecast may give the pollutants concentration or the air quality index.

Countries and cities are given forecasts by state and local government organizations, as well as private companies like Airly, AirVisual, Aerostate, BreezoMeter, PlumeLabs, and DRAXIS that give air pollution forecast.

### Techniques

- Air pollution forecasting can be done by coupling weather forecasting systems with Chemical transport model and Atmospheric dispersion modeling.

- Using machine learning techniques.

The forecast takes into account local emission sources (like nearby traffic or industry) and remote sources (e.g. dust that is carried by air parcels and follows the wind direction).

The forecast temporally resolution is usually daily or hourly and the spatial resolution can change from block resolution to dozens of km resolution.

Most forecasts of air quality cover two to five days.

### Motivation

- By knowing the air quality forecast one can decide how to act, e.g. due to air pollution health effects, one can choose the best time to do an outdoor activity.

- Deciding whether to put on a skin care ointment.

- Find the cleanest route for cars.

- Deciding whether to leave the windows open or closed.

## Air Pollution Hotspots

Air pollution hotspots are areas where air pollution emissions expose individuals to increased negative health effects. Hotspots denote areas in which a population's exposure to pollution and estimated health risks are high. Air pollution hotspots are particularly common in highly populated, urban areas, where there may a combination of stationary sources (e.g. industrial facilities) and mobile sources (e.g. cars and trucks) of pollution. Emissions from these sources can cause respiratory disease, childhood asthma, cancer, and other health problems. A fine particulate matter such as diesel soot, which contributes to more than 3.2 million premature deaths around the world each year, is a significant problem. It is very small and can lodge itself within the lungs and enter the bloodstream. Diesel soot is concentrated in densely populated areas, and one in six people in the U.S. live near a diesel pollution hot spot.

While air pollution hotspots affect a variety of populations, some groups are more likely to be located in hotspots. Previous studies have shown disparities in exposure to pollution by race and/or income (cite one of the early readings from our syllabus, e.g. Mohai & Pellow, or Saha). Hazardous land uses (toxic storage and disposal facilities, manufacturing facilities, major roadways) tend to be located where property values and income levels are low. Low socioeconomic status can be a proxy for other kinds of social vulnerability, including race, a lack of ability to influence regulatory permitting and a lack of ability to move to neighborhoods with less environmental pollution. These communities bear a disproportionate burden of environmental pollution and are more likely to face health risks such as cancer or asthma.

Studies show that patterns in race and income disparities not only indicate a higher exposure to pollution but also higher risk of adverse health outcomes. Communities characterized by low socioeconomic status and racial minorities can be more vulnerable to cumulative adverse health impacts resulting from elevated exposure to pollutants than more privileged communities. Blacks and Latinos generally face more pollution than whites and Asians, and low-income communities bear a higher burden of risk than affluent ones. Racial discrepancies are particularly distinct in suburban areas of the South and metropolitan areas of the West. Residents in public housing, who are generally low-income with poor access to health care and cannot move to healthier neighborhoods, are highly affected by nearby refineries and chemical plants.

Community groups and academic researchers have argued the unequal distribution of pollution on the poor and communities of color is an "environmental justice".

Policy makers and researchers concerned with improving environmental justice for communities situated next to major sources of air pollution have developed a number of regulatory tools to identify air pollution hotspots. The EPA, for example, utilizes their Risk-Screening Environmental Indictors (RSEI) model to identify hotspots from a score of 3 to 15, with higher scores indicating closer proximity to hazards. Individual states have also taken steps to improve identification and surveillance. California's AB 2588 Air Toxics "Hot Spots" Program, enacted in 1987, seeks to collect emission data, determine health risks, and notify local residents of major risks. By identifying hotspots regulators hope these tools will help them reduce pollution and inform nearby populations through the health risk assessments of individual pollutants and facilities that are summed in each zone to develop a total lifetime cancer risk. Air pollution hot spots are also at issue in pollution-trading programs, such as cap-and-trade systems designed to control pollution. These programs can potentially exacerbate effects from air pollution hotspots if the differences in chemical hazards are ignored. These programs also cause pollution to be mitigated towards where credit-buying firms are located. Factories can purchase emissions reduction credits from other firms, which leads to concentrated areas of pollution since facilities that sell their credits are "exporting" their pollution to firms more likely to buy credits. However, some studies have noted that these claims have not materialized. Evan Ringquist, a professor at Indiana University of Public and Environmental Affairs, states that there is little empirical evidence to suggest the emergence of hotspots.A Cedars-Sinai study found that prolonged exposure to particulate matter in air pollution in the Los Angeles Basin triggered inflammation and the appearance of cancer-related genes in the brains of rats.

## Lead Pollution

Sources of lead emissions vary from one area to another. At the national level, major sources of lead in the air are ore and metals processing and piston-engine aircraft operating on leaded aviation fuel. Other sources are waste incinerators, utilities, and lead-acid battery manufacturers. The highest air concentrations of lead are usually found near lead smelters.

As a result of EPA's regulatory efforts including the removal of lead from motor vehicle gasoline, levels of lead in the air decreased by 98 percent between 1980 and 2014.

### Effects of Lead on Human Health

Once taken into the body, lead distributes throughout the body in the blood and is accumulated in the bones. Depending on the level of exposure, lead can adversely affect the nervous system, kidney function, immune system, reproductive and developmental systems and the cardiovascular system. Lead exposure also affects the oxygen carrying capacity of the blood. The lead effects most commonly encountered in current populations are neurological effects in children and cardiovascular effects (e.g., high blood pressure and heart disease) in adults. Infants and young children are especially sensitive to even low levels of lead, which may contribute to behavioral problems, learning deficits and lowered IQ.

## Effects of Lead on Ecosystems

Lead is persistent in the environment and can be added to soils and sediments through deposition from sources of lead air pollution. Other sources of lead to ecosystems include direct discharge of waste streams to water bodies and mining. Elevated lead in the environment can result in decreased growth and reproductive rates in plants and animals, and neurological effects in vertebrates.

## References

- Mcdougal, Myres S. And Schlei, Norbert A. "The Hydrogen Bomb Tests in Perspective: Lawful Measures for Security". In Myres S. Mcdougal, et al. (1987), Studies in World Public Order, p. 766. New Haven: New Haven Press. ISBN 0-89838-900-3

- Air-pollution, entry: newworldencyclopedia.org, Retrieved 18 April, 2019

- Johnson, Carl (1984). "Cancer Incidence in an Area of Radioactive Fallout Downwind From the Nevada Test Site". Journal of the American Medical Association. 251 (2): 230. Doi:10.1001/jama.1984.03340260034023

- Smogpollution: conserve-energy-future.com, Retrieved 21 July, 2019

- "Dermalogica & breezometer partner to educate on pollution's skin effects". Retrieved 31 May 2018

- Mohai, P; Lantz, PM; Morenoff, J; House, JS; Mero, RP (2009). "Racial and Socioeocnomic Disparities in Residential Proximity". American Journal of Public Health. 99 (3): S649–S656. Doi:10.2105/ajph.2007.131383. PMC 2774179. PMID 19890171

- Gerrard, Michael B.; Sheila R. Foster, eds. (2008). Law of environmental justice : theories and procedures to address disproportionate risks (2nd ed.). Chicago, Ill.: American Bar Association, Section of Environment, Energy, and Resources. ISBN 978-1604420838

- Basic-information-about-lead-air-pollution, lead-air-pollution: epa.gov, Retrieved 13 July, 2019

# 2

# Air Pollutants

Some of the different types of air pollutants are toxic air pollutants, criteria air pollutants and indoor air pollutants. A few examples of air pollutants are sulfur dioxide, nitrogen oxide, ozone and carbon monoxide. This chapter has been carefully written to provide an easy understanding of these types of air pollutants and air pollution dispersion model.

## Toxic Air Pollutants

Toxic, or hazardous, air pollutants cause or are suspected of causing cancer, birth defects, or other serious harms. They can be gases, like hydrogen chloride, benzene or toluene, dioxin, or compounds like asbestos, or elements such as cadmium, mercury, and chromium. The U.S. Environmental Protection Agency has classified 187 pollutants as hazardous.

Just because a pollutant is not listed on the list as "hazardous" does not mean that it does not cause cancer or is safe to breathe, however. Other air pollutants like particle pollution can also cause cancer or other serious hazards.

### Health Effects from Toxic Air Pollutants

Toxic air pollutants pose different risks to health depending on the specific pollutant, including:

- Cancer, including lung, kidney, bone, stomach.

- Harm to the nervous system and brain.

- Birth defects.

- Irritation to the eyes, nose and throat.

- Coughing and wheezing.

- Impaired lung function.

- Harm to the cardiovascular system.

- Reduced fertility.

## How are People Exposed to these Pollutants?

People inhale many of these pollutants in the air where they live. But, since these pollutants also settle into waterways, streams, rivers and lakes, people can drink them in the water or eat them in the fish from these waters. Some hazardous pollutants settle into the dirt that children play in and may put in their mouths.

## Where do Toxic air Pollutants Come from?

Major sources of toxic air pollutants outdoors include emissions from coal-fired power plants, industries, and refineries, as well as from cars, trucks and buses.

Indoor air also can contain hazardous air pollutants from sources that include tobacco smoke, building materials like asbestos, and chemicals like solvents.

# Criteria Air Pollutants

Criteria air Pollutants (CAP), or criteria pollutants, are a set of air pollutants that cause smog, acid rain, and other health hazards. CAPs are typically emitted from many sources in industry, mining, transportation, electricity generation and agriculture. In many cases they are the products of the combustion of fossil fuels or industrial processes.

The six criteria air pollutants were the first set of pollutants recognized by the United States Environmental Protection Agency as needing standards on a national level. The Clean Air Act requires the EPA to set US National Ambient Air Quality Standards (NAAQS) for the six CAPs. The NAAQS are health based and the EPA sets two types of standards: primary and secondary. The primary standards are designed to protect the health of 'sensitive' populations such as asthmatics, children, and the elderly. The secondary standards are concerned with protecting the environment. They are designed to address visibility, damage to crops, vegetation, buildings, and animals.

## EPA Endangerment Findings/ Mass v. EPA

In 2009, the EPA Administrator found that under section 202(a) of the Clean Air Act greenhouse gases threaten both the public health and the public welfare, and that greenhouse gas emissions from motor vehicles contribute to that threat. This final action has two distinct 'findings,' which are:

1. The 'Endangerment Finding' in which the Administrator found that the mix of atmospheric concentrations of six key, well-mixed greenhouse gases threatens both the public health and the public welfare of current and future generations. These six greenhouse gases are: carbon dioxide ($CO_2$), methane ($CH_4$), nitrous oxide ($N_2O$), hydrofluorocarbons (HFCs), perfluorocarbons (PFCs), and sulfur hexafluoride ($SF_6$). These greenhouse gases in the atmosphere constitute the "air pollution" that threatens both public health and welfare.

2. The 'Cause or Contribute Finding,' in which the Administrator found that the combined greenhouse gas emissions from new motor vehicles and motor vehicle engines contribute to the atmospheric concentrations of these key greenhouse gases and hence to the threat of climate change.

The EPA issued these endangerment findings in response to the 2007 supreme court case Massachusetts v. EPA, when the court determined that greenhouse gases are air pollutants according to the Clean Air Act. The court made the decision that the EPA must determine whether greenhouse gas emissions from new motor vehicles "cause or contribute to air pollution which may be reasonably be anticipated to endanger public health or welfare, or whether the science is too uncertain to make a reasoned decision" (EPA's Endangerment Finding).

The EPA determined that, according to this decision, there are six greenhouse gases that need to be regulated. These include:

- Carbon dioxide ($CO_2$)

- Methane ($CH_4$)

- Nitrous oxide ($N_2O$)

- Hydrofluorocarbons ($HFC_s$)

- Perfluorocarbons ($PFC_s$)

- Sulfur hexafluoride ($SF_6$)

This action allowed the EPA to set the greenhouse gas emission standards to light-duty vehicles proposed jointly with the Department of Transportation's Corporate Average Fuel Economy (CAFE) standards in 2009.

## Petition to Add Seven Criteria Air Pollutants

On December 2, 2009, the Center for Biological Diversity and 350.org requested that the EPA recognize that carbon dioxide and other GHGs are reasonably anticipated to endanger public health and welfare. Petitioners proposed that EPA list carbon dioxide as a criteria air pollutant, as outlined in the Clean Air Act. They also requested that the EPA set NAAQS for carbon dioxide at no greater than 350 ppm- a "level that accurately reflects the most recent scientific knowledge." Petitioners further requested that EPA designate the five other greenhouse gases, highlighted in Mass v. EPA, as criteria pollutants as well and establish pollution caps for them. Furthermore, the petitioners proposed that nitrogen trifluoride ($NF_3$) be regulated as a criteria air pollutant in addition to the other six.

# Sulfur Dioxide

## Emission Sources and Trends

The main sources of Sulphur dioxide ($SO_2$) emissions are electricity generation, industrial and domestic fuel combustion. Total $SO_2$ emissions have decreased substantially, in line with changes in fuel use and commitments to international agreements within the UNECE Convention on Long Range Transboundary Air Pollution (CLRTAP) and the European Union ("Sulphur Protocols"). The UK is also a signatory to the Gothenburg Protocol. This commits the UK to reducing emissions of S to 625 Gg-$SO_2$ (313 Gg-S) by 2010. In addition, the UK is also a signatory to the EU National Emission Ceilings Directive, which requires 2010 emissions of $SO_2$ to be 585 Gg (293 Gg-S) or lower.

Heavy industry declined from the late 1970s through to the 1990s as the UK underwent a change from an economy dominated by manufacturing to a service-based economy. As a result there were substantial reductions in the levels of heavy industry in the UK, most notably in the iron and steel sector. This gives rise to the emission reductions observed in the 1970s. Substantial reductions were seen from the early1990s onwards as there was a switch from coal to gas in the domestic, industrial and electricity generating sectors. In 1992 the use of gas was responsible for only 2% of electricity generation, but by 2000 this had grown to 34%.

Emissions of sulphur still show a steady decline due to legislation associated with restricting sulphur emissions from large combustion plants, which has resulted in the installation and use of emissions abatement equipment. Flue gas desulphurisation abatement equipment is highly effective at removing suplhur from stack emissions, and its use has ensured that emissions have continued to fall, even though coal use has increased since 2000.

Sulphur emissions from shipping - International shipping is a large source of sulphur and activity levels of shipping in general have been increasing with time as other sources have been significantly reduced. As a result, current trends suggest that shipping is set to become one of the most important contributors to UK sulphur emissions in future years.

## Impacts and Recovery

Background level concentrations of $SO_2$ in the UK have fallen so much that there is no longer a threat to plant health.

The UK has one of the most diverse lichen flora in the world. Historical records and field observations clearly document the widespread loss of lichen species across many areas of England from the last century, with air pollution levels and changes in land use being the most likely causes. There are large differences in the extent of the loss of different species, largely reflecting their sensitivity to $SO_2$; indeed, methods were developed of mapping $SO_2$ concentrations using the range of lichen species found at different sites.

Decreases in $SO_2$ concentrations over the past two to three decades have had detectable effects on vegetation, including substantial increases in the distribution of many lichen species, improved tree growth in certain areas and increased likelihood of sulphur deficiency in crops. Rose & Hawksworth (1981) examined lichen populations in north and west London, and found several species, such as Evernia prunasti, Parmelia caperata, Parmelia subaurifera and Usnea subfloridana, that were rare or extinct in the area prior to 1970.

# Nitrogen Oxide

Nitrogen dioxide is an irritant gas, which at high concentrations causes inflammation of the airways.

When nitrogen is released during fuel combustion it combines with oxygen atoms to create nitric oxide (NO). This further combines with oxygen to create nitrogen dioxide ($NO_2$). Nitric oxide is not considered to be hazardous to health at typical ambient concentrations, but nitrogen dioxide can be. Nitrogen dioxide and nitric oxide are referred to together as oxides of nitrogen ($NO_x$).

$NO_x$ gases react to form smog and acid rain as well as being central to the formation of fine particles (PM) and ground level ozone, both of which are associated with adverse health effects.

## Sources of $NO_x$ Pollution

$NO_x$ is produced from the reaction of nitrogen and oxygen gases in the air during combustion, especially at high temperatures. In areas of high motor vehicle traffic, such as in large cities, the amount of nitrogen oxides emitted into the atmosphere as air pollution can be significant. $NO_x$ gases are formed whenever combustion occurs in the presence of nitrogen – e.g. in car engines; they are also produced naturally by lightning.

## Health Issues Created by $NO_x$

$NO_x$ mainly impacts on respiratory conditions causing inflammation of the airways at high levels. Long term exposure can decrease lung function, increase the risk of respiratory conditions and increases the response to allergens. $NO_x$ also contributes to the formation of fine particles (PM) and ground level ozone, both of which are associated with adverse health effects.

## Impact of Nitrogen Dioxide on Ecosystems

High levels of $NO_x$ can have a negative effect on vegetation, including leaf damage and reduced growth. It can make vegetation more susceptible to disease and frost damage. A study of the effect of nitrogen dioxide and ammonia ($NH_3$) on the habitat of Epping Forest has revealed that pollution is likely to be significantly influencing ecosystem health in the forest. The study demonstrated that local traffic emissions contribute substantially to exceeding the critical levels and critical loads in the area. The critical level for the protection of vegetation is 30 µg/m$_3$ measured as an annual average.

NOx also reacts with other pollutants in the presence of sunlight to form ozone which can damage vegetation at high concentrations.

Critical Level is the threshold level for the atmospheric concentration of a pollutant above which harmful direct effects can be shown on a habitat or species. Critical Load is the threshold level for the deposition of a pollutant above which harmful indirect effects can be shown on a habitat or species.

## $NO_x$ Level Objectives

Twelve European Member States exceeded one or more of the emission limits set by the EU National Emission Ceilings (NEC) Directive, according to recent official data for 2010 reported to the European Environment Agency (EEA). In some instances the limits were exceeded by significant amounts. The pollutant for which most exceedances were registered was $NO_x$.

The European Union sets Limit Values for a range of pollutants that are considered to be harmful to health and the environment. The European Commission can take action against any Member State if the air quality does not meet the Limit Values throughout its territory.

## Ozone

Ozone ($O_3$) is present throughout the atmosphere although there are concentration peaks at two levels, the stratosphere (15 - 50 km) and troposphere (0-15 km), with the largest fraction and concentrations being in the stratospheric $O_3$ layer. Stratospheric $O_3$ is important as it regulates the transmittance of ultraviolet light to the surface of the earth. Hence reductions in stratospheric $O_3$ in polar regions, particularly the Antarctic "ozone hole", are of concern regarding the health effects of exposure to increased levels of UV-B.

In contrast, $O_3$ in the troposphere (ground level) is regionally important as a toxic air pollutant and greenhouse gas. Mixing with stratospheric air provides a natural global average background of around 10-20 parts per billion (ppb), thqough there is some debate about the concentration. Additional quantities of tropospheric $O_3$ are produced by photochemical reactions from nitrogen oxides ($NO_x$) and volatile organic compounds (VOCs), which include various hydrocarbons.

## Formation and Sources

Ground-level ozone ($O_3$) is not emitted directly from anthropogenic sources. It is a "secondary" pollutant formed by a complicated series of chemical reactions in the presence of sunlight. Photochemical reactions of $NO_x$ and VOCs (originating from largely from combustion processes) govern the concentration of ground-level $O_3$ in the atmosphere. Under typical daytime conditions with a well-mixed atmosphere, three reactions reach equilibrium and no net chemistry occurs.

$$NO + O_3 \rightarrow NO_2 + O_2$$

$$NO_2 + hn \rightarrow NO + O$$

$$O + O_2 \rightarrow O_3 (+M)$$

where, hn = sunlight with wavelength 280-430 nm, M = any molecule eg $N_2$ or $O_2$.

The chemical reactions do not take place instantaneously, but can take hours or days. Ozone levels at a particular location may have arisen from VOC and $NO_x$ emissions many hundreds or even thousands of miles away. Maximum concentrations, therefore, generally occur downwind of the source areas of the precursor pollutant emissions.

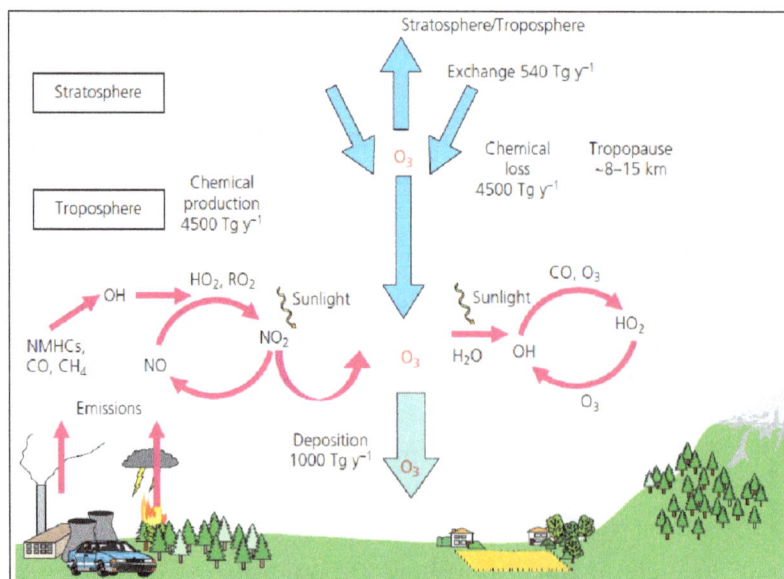

A schematic view of the sources and sinks of $O_3$ in the troposphere.

Annual global fluxes of $O_3$ calculated using a global chemistry–transport model have been included to show the magnitudes of the individual terms. These fluxes include stratosphere to troposphere exchange, chemical production and loss in the troposphere and the deposition flux to terrestrial and marine surfaces.

Anthropogenic emissions of the ozone precursors ($NO_x$/VOCs) can also cause large transient increases in ozone concentration, termed episodes or smog. These occur when high concentrations of precursors coincide with weather conditions favourable for ozone production such as when the air is warm and slow moving. These "ozone episodes" provide concentrations of $O_3$ (>40 ppb) which are toxic both to human health (with a long term threshold objective of 50 ppb daily 8-h mean, WHO) and vegetation.

Prior to the industrial revolution natural sources of $NO_x$ and VOCs would have generated $O_3$ in the troposphere, adding to that transported from the stratosphere. However, the large amounts of $NO_x$ and VOCs released by human activities, such as the combustion of fossil fuels, has led to a large increase in the northern hemisphere background concentration. Evaluation of historical $O_3$ measurements indicate that since the 1950s, the background ozone concentration has roughly doubled, although there has been some slowing down of this trend in the last decade.

Table: Summary of Ozone formation from natural and anthropogenic $NO_x$ and VOC sources.

| $NO_x$ | | VOC | |
|---|---|---|---|
| Natural | Anthropogenic | Natural | Anthropogenic |
| Soils, natural fires | Transport (road, sea and rail), power stations, other industry and combustion processes | Vegetation, natural fires | Transport, combustion processes, solvents, oil production |

## Concentrations

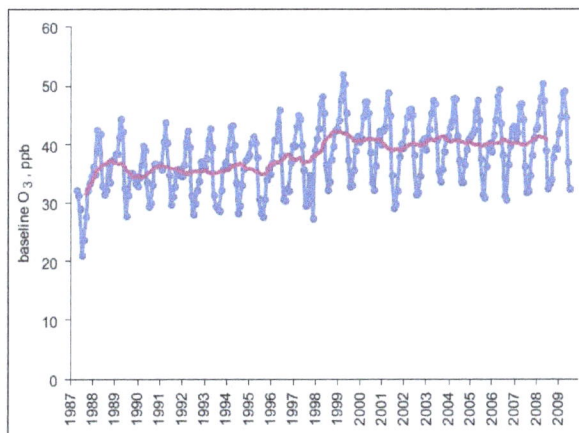

Trends in monthly mean baseline ozone levels at Mace Head on the west coast of Ireland between 1986 and 2006, including a running 12-month mean (pink line), showing the ~5 ppb increase in ozone mixing ratios over 20 years.

Policy actions to date across Europe have reduced the emissions of ozone precursors $NO_x$ and VOCs. These emission controls have reduced peak ozone concentrations by typically 30 ppb in the UK, but over the last 20 years mean concentrations have been increasing in urban areas due to reductions in the local depletion of $O_3$ by NO. Background concentrations have increased in rural areas due to increases in the hemispheric background $O_3$ concentrations which have increased by approximately 0.2 ppb per year, or by about 5 ppb. The cause of the increases in background $O_3$ are increases in precursor emissions throughout the northern hemisphere, including shipping, aircraft, vehicle, and industrial emissions in developing economies.

Ozone concentrations are highly variable, spatially and temporally. Figure shows the high background concentration in March to May over the northern half of Scotland, upland Wales and northern England, and parts of southern and eastern England. In the summer months of May-July high concentrations cover parts of south-east England and some upland areas of Wales and northern Scotland. The summer time spatial pattern extends across Western Europe and is caused by regions where ozone production occurs more frequently in a combination of precursor emissions ($NO_x$ and VOCs), high solar radiation and temperatures. During the winter the highest concentrations occurring in the NW and the lowest in the SE as a consequence of ozone destruction in the NO polluted atmosphere of the dense urban/industrial regions of the southern UK and continental Western Europe.

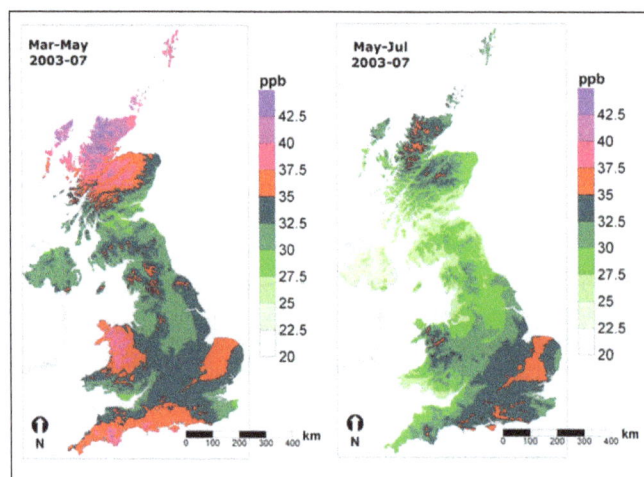

Distribution of mean ozone concentrations over the periods March-May and May-July, using data for the five year period 2003-07.

## Effects

A large body of evidence has shown that ambient $O_3$ causes damage to $O_3$-sensitive vegetation. $O_3$ enters leaves via the stomatal pores on the leaf surface. Once inside the leaf, a series of chemical reactions occur leading to damage to cell membranes and other negative impacts on plant metabolism, including photosynthesis. These effects can be in response to short-term episodes or cumulative during the growing season, and can lead to:

- Visible leaf damage and premature aging of leaves;

- Reductions in above- and below-ground growth and biomass;

- Changes in the ratio between shoot and root biomass (including carbon allocation);

- Reductions in flower number, flower biomass and seed production;

- Reductions in crop yield quantity and quality, including cereal grains, potato tubers and tomato fruit;

- Changes in forage quantity and quality for pasture;

- Altered tolerance to abiotic stresses such as drought and frost and biotic stresses such as pest attacks and diseases.

## Critical Levels

Ozone pollution effects on vegetation are described by cumulative metrics that are based on either atmospheric concentration of $O_3$ over a threshold concentration (AOT40) or modelled uptake of $O_3$ through the stomatal pores on the leaf surface (Phytotoxic Ozone Dose above a threshold of Y, $POD_Y$). A recent analysis of field evidence for $O_3$ effects confirmed that maps generated using an $O_3$ flux metric more closely matched locations of damage than those based on $O_3$ concentration. New $O_3$ flux-based critical levels have been derived for crops, tree species and for grassland species.

## Other Impacts

The role of ozone as a contributor to the direct radiative forcing of global climate has grown in importance. In addition, the recognition that the effects of $O_3$ on carbon sequestration through its effects on primary productivity of vegetation is an additional reason for interest in ground level ozone.

## Tropospheric Ozone

Ozone ($O_3$) is a trace gas of the troposphere, with an average concentration of 20-30 parts per billion by volume (ppbv), with close to 100 ppbv in polluted areas. Ozone is also an important constituent of the stratosphere, where the ozone layer exists. The troposphere is the lowest layer of the Earth's atmosphere. It extends from the ground up to a variable height of approximately 14 kilometers above sea level. Ozone is least concentrated in the ground layer (or planetary boundary layer) of the troposphere. Its concentration increases as height above sea level increases, with a maximum concentration at the tropopause. About 90% of total ozone in the atmosphere is in the stratosphere, and 10% is in the troposphere. Although tropospheric ozone is less concentrated than stratospheric ozone, it is of concern because of its health effects. Ozone in the troposphere is considered a greenhouse gas, and may contribute to global warming.

Photo-chemical and chemical reactions involving ozone drive many of the chemical processes that occur in the troposphere by day and by night. At abnormally high concentrations (the largest source being emissions from combustion of fossil fuels), it is a pollutant, and a constituent of smog.

Photolysis of ozone occurs at wavelengths below approximately 310-320 nano-meters. This reaction initiates the chain of chemical reactions that remove carbon monoxide, methane, and other hydrocarbons from the atmosphere via oxidation. Therefore, the concentration of tropospheric ozone affects how long these compounds remain in the air. If the oxidation of carbon monoxide or methane occur in the presence of nitrogen monoxide (NO), this chain of reactions has a net product of ozone added to the system.

## Measurement

Ozone in the atmosphere can be measured by remote sensing technology, or by *in-situ* monitoring technology. Because ozone absorbs light in the UV spectrum, the most common way to measure ozone is to measure how much of this light spectrum is absorbed in the atmosphere. Because the stratosphere has higher ozone concentration than the troposphere, it is important for remote sensing instruments to be able to determine altitude along with the concentration measurements. The TOMS-EP instrument aboard a satellite from NASA is an example of an ozone layer measuring satellite, and TES is an example of an ozone measuring satellite that is specifically for the troposphere. Lidar is a common ground based remote sensing technique to measure ozone. TOLnet is the network of ozone observing lidars across the United States.

Ozonesondes are a form of in situ, or local measurements. An ozonesonde is an ozone measuring instrument attached to a meteorological balloon, so that the instrument can directly measure ozone concentration at the varying altitudes along the balloon's upward path. The information collected from the instrument attached to the balloon is transmitted back using radiosonde technology. NOAA has worked to create a global network of tropospheric ozone measurements using ozonesondes.

Ozone is also measured in air quality environmental monitoring networks. In these networks, in-situ ozone monitors based on ozone's UV-absorption properties are used to measure ppb-levels in ambient air.

## Formation

The majority of tropospheric ozone formation occurs when nitrogen oxides ($NO_x$), carbon monoxide (CO) and volatile organic compounds (VOCs), react in the atmosphere in the presence of sunlight, specifically the UV spectrum. $NO_x$, CO, and VOCs are considered ozone precursors. Motor vehicle exhaust, industrial emissions, and chemical solvents are the major anthropogenic sources of these ozone precursors. Although the ozone precursors often originate in urban areas, winds can carry $NO_x$ hundreds of kilometers, causing ozone formation to occur in less populated regions as well.

The chemical reactions that produce tropospheric ozone are a series of interrelated cycles (known as the HOx and $NO_x$ cycles); They start with the oxidation of carbon monoxide (CO) or VOCs (such as butane). To begin the process, CO and VOCs are oxidized by the hydroxyl radical ($\cdot OH$) to form carbon dioxide ($CO_2$), and water ($H_2O$) in the CO oxidation case. These oxidizing reactions then produce the peroxy radical ($HO_2\cdot$) that will react with NO to produce $NO_2$. $NO_2$ is subsequently photolyzed during by daytime, thus resulting in NO and a single oxygen atom. This single oxygen atom reacts with molecular oxygen $O_2$ to produce ozone.

An outline of the chain reaction that occurs in oxidation of CO, producing $O_3$:

The reaction begins with the oxidation of CO by the hydroxyl radical ($\cdot OH$). The radical adduct ($\cdot HOCO$) is unstable and reacts rapidly with oxygen to give a peroxy radical, $HO_2\cdot$:

$$\cdot OH + CO \rightarrow \cdot HOCO$$

$$\cdot HOCO + O_2 \rightarrow HO_2\cdot + CO_2$$

Peroxy-radicals then go on to react with NO to produce $NO_2$, which is photolysed by UV-A radiation to give a ground-state atomic oxygen, which then reacts with molecular oxygen to form ozone.

$$HO_2\cdot + NO \rightarrow \cdot OH + NO_2$$

$$NO_2 + h\nu \rightarrow NO + O(^3P), \lambda < 400 \text{ nm}$$

$$O(^3P) + O_2 \rightarrow O_3$$

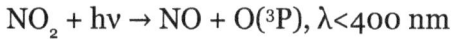

Note that these three reactions are what forms the ozone molecule, and will occur the same way in the oxidation of CO or VOCs case.

The amount of ozone produced through these reactions in ambient air can be estimated using a modified Leighton relationship. The limit on these interrelated cycles producing ozone is the reaction of $\cdot OH$ with $NO_2$ to form nitric acid at high $NO_x$ levels. If nitrogen monoxide (NO) is instead present at very low levels in the atmosphere (less than 10 approximately ppt), the of peroxy radicals ($HO_2 \cdot$) formed from the oxidation will instead react with themselves to form peroxides, and not produce ozone.

## Health Effects

Health effects depend on ozone precursors, which is a group of pollutants, primarily generated during the combustion of fossil fuels. Reaction with daylight ultraviolet (UV) rays and these precursors create ground-level ozone pollution (Tropospheric Ozone). Ozone is known to have the following health effects at concentrations common in urban air:

- Irritation of the respiratory system, causing coughing, throat irritation, and/or an uncomfortable sensation in the chest.

- Reduced lung function, making it more difficult to breathe deeply and vigorously. Breathing may become more rapid and more shallow than normal, and a person's ability to engage in vigorous activities may be limited.

- Aggravation of asthma. When ozone levels are high, more people with asthma have attacks that require a doctor's attention or use of medication. One reason this happens is that ozone makes people more sensitive to allergens, which in turn trigger asthma attacks.

- Increased susceptibility to respiratory infections.

- Inflammation and damage to the lining of the lungs. Within a few days, the damaged cells are shed and replaced much like the skin peels after a sunburn. Animal studies suggest that if this type of inflammation happens repeatedly over a long time period (months, years, a lifetime), lung tissue may become permanently scarred, resulting in permanent loss of lung function and a lower quality of life.

A statistical study of 95 large urban communities in the United States found significant association between ozone levels and premature death. The study estimated that a one-third reduction in urban ozone concentrations would save roughly 4000 lives per year (Bell et al., 2004). Tropospheric Ozone causes approximately 22,000 premature deaths per year in 25 countries in the European Union.

## Problem Areas

The United States Environmental Protection Agency has developed an Air Quality index to help explain air pollution levels to the general public. 8-hour average ozone mole fractions of 76 to 95 nmol/mol are described as "Unhealthy for Sensitive Groups", 96 nmol/mol to 115 nmol/mol as "unhealthy" and 116 nmol/mol to 404 nmol/mol as "very unhealthy". The EPA has designated over

300 counties of the United States, clustered around the most heavily populated areas (especially in California and the Northeast), as failing to comply with the National Ambient Air Quality Standards.

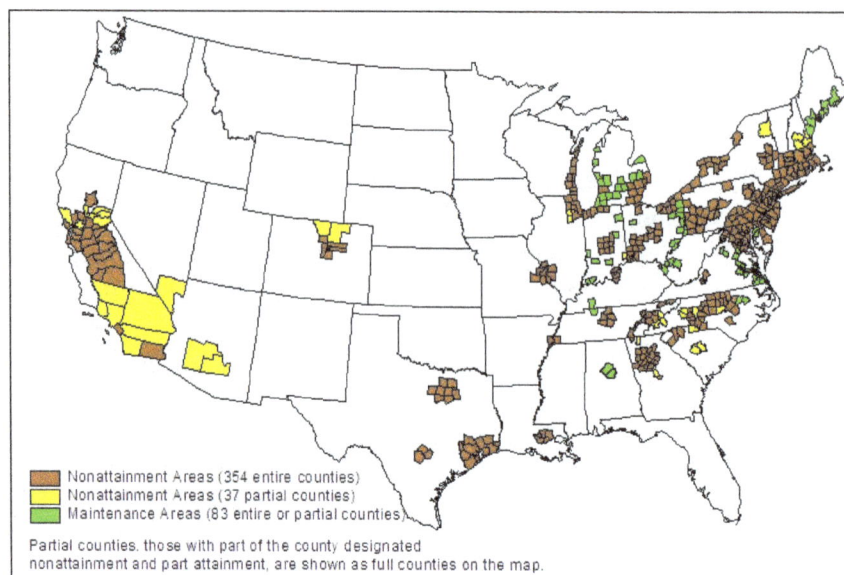

Nonattainment Areas (354 entire counties)
Nonattainment Areas (37 partial counties)
Maintenance Areas (83 entire or partial counties)

Partial counties, those with part of the county designated nonattainment and part attainment, are shown as full counties on the map.

## Climate Change

Melting of sea ice releases molecular chlorine, which reacts with UV radiation to produce chlorine radicals. Because chlorine radicals are highly reactive, they can expedite the degradation of methane and tropospheric ozone and the oxidation of mercury to more toxic forms. Ozone production rises during heat waves, because plants absorb less ozone. It is estimated that curtailed ozone absorption by plants is responsible for the loss of 460 lives in the UK in the hot summer of 2006. A similar investigation to assess the joint effects of ozone and heat during the European heat waves in 2003, concluded that these appear to be additive.

# Carbon Monoxide

Carbon monoxide (CO)—a colorless, odorless, tasteless, and toxic air pollutant—is produced in the incomplete combustion of carbon-containing fuels, such as gasoline, natural gas, oil, coal, and wood. The largest anthropogenic source of CO in the United States is vehicle emissions. Breathing the high concentrations of CO typical of a polluted environment leads to reduced oxygen ($O_2$) transport by hemoglobin and has health effects that include headaches, increased risk of chest pain for persons with heart disease, and impaired reaction timing. In the 1960s, vehicle emissions led to increased and unhealthful ambient CO concentrations in many U.S. cities. With the introduction of emissions controls, particularly automotive catalysts, estimated CO emissions from all sources decreased by 21% from 1980 to 1999. Average ambient concentrations decreased by about 57% over the same period.

The locations that continue to have high concentrations of CO tend to have topographical or meteorological characteristics that exacerbate pollution; for example, strong temperature inversions or the existence of nearby hills that inhibit wind flow may limit pollutant dispersion. Because of the limited dispersion, many of those areas also have unhealthful concentrations of summer ozone ($O_3$) and year-round particulate matter (PM). Low temperatures also contribute to high CO concentrations. Engines and vehicle emissions-control equipment operate less efficiently when cold: Air-to-fuel ratios are lower, combustion is less complete, and catalysts take longer to become fully operational. The result is that products of incomplete combustion, including CO, are formed in higher concentrations. Sometimes, topography, meteorology, and emissions combine to cause high concentrations of CO. Compliance with the health-based National Ambient Air Quality Standards (NAAQS) for CO has proved difficult under those circumstances. The question arises as to whether unique methods are necessary to manage CO in such problem areas, or whether the current policies will ultimately achieve good air quality.

## Health Effects of CO

## Clinical and Epidemiological Studies of CO Effects

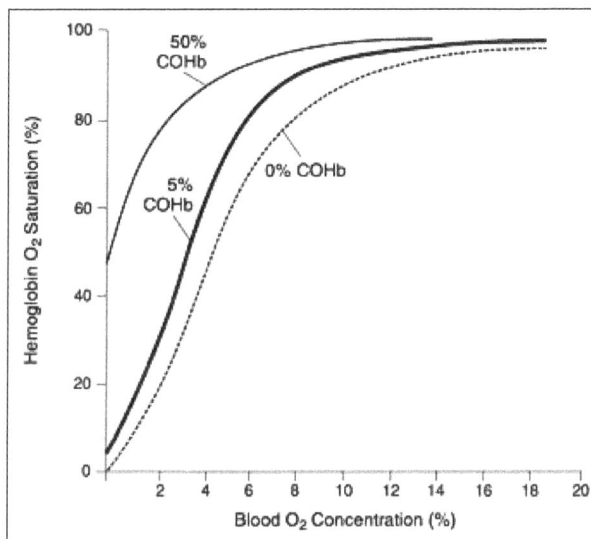

Diagram of hemoglobin response to the presence of COHb.

The concentration of $O_2$ in the environment surrounding the hemoglobin is shown on the x-axis. The $O_2$ saturation, or how much of the hemoglobin's capacity for storing $O_2$ is used, is shown on the y-axis. At higher $O_2$ concentrations, as are found in the lungs, the hemoglobin can be more $O_2$ saturated. Likewise, at lower $O_2$ concentrations, as are found in other parts of the body, $O_2$ will dissociate from the hemoglobin to achieve $O_2$ saturations as indicated by the curve. The presence of COHb shifts this curve to the left. For a given $O_2$ concentration, the hemoglobin will require a higher $O_2$ saturation and allow less $O_2$ to be released to body's tissues.

CO affects human health by impairing the ability of the blood to bring $O_2$ to body tissues. When CO is inhaled, it rapidly crosses the alveolar epithelium to reach the blood, where it binds to hemoglobin to form carboxyhemoglobin (COHb), a useful marker for predicting the health effects of CO. Because CO has an affinity for hemoglobin more than 200 times greater than does $O_2$, the presence of CO in the lung will displace $O_2$ from the hemoglobin. In other words, when CO is present in the lungs, the hemoglobin will be unable to reach 100% $O_2$ saturation. In addition, the presence of COHb increases hemoglobin's affinity for $O_2$, thereby inhibiting release of $O_2$ from the hemoglobin to body tissues. The effect of COHb is illustrated by a leftward shift of the $O_2$-hemoglobin dissociation curve. As shown in figure, once COHb forms, the hemoglobin is unable to reach 0% $O_2$ saturation. This second effect continues until the COHb dissociates, typically several hours after CO exposure. CO not only decreases the $O_2$-carrying capacity of the blood, but also decreases the ability of the tissues to extract $O_2$ from the blood during circulation. CO has also been shown to bind to myoglobin and may affect $O_2$ transport to muscle.

COHb levels in healthy individuals not recently exposed to high concentrations of ambient CO are 0.3 to 0.7%. Exposure to high concentrations of ambient CO can result in concentrations of COHb of 2% or higher if the exposure lasts long enough (hours). For those who smoke, cigarette-smoking is typically the most significant source of personal CO exposure. COHb concentrations, which are generally less than 1% in nonsmokers, average about 5% in smokers and are up to 10% or even higher in some very heavy smokers.

The CO health standards set by EPA are intended to keep COHb concentrations for nonsmokers below 2% in order to protect the most susceptible members of the population. EPA last published the Air Quality Criteria for Carbon Monoxide, which summarizes health research findings and their implications for setting the NAAQS, in June 2000. That document provides a comprehensive review of the literature pertaining to the health effects of CO for typical environmental exposures that would be associated with COHb levels less than 10%. The major findings presented by EPA are summarized below; the committee accepts these findings as sufficient evidence of the health effects caused by exposure to CO at concentrations of 9 parts per million (ppm) and above for an extended period of time.

The acute affects of CO poisoning are well understood. Generally, in otherwise healthy people, headache develops when COHb concentrations reach 10%; tinnitus (ringing in the ear) and light-headedness at 20%; nausea, vomiting, and weakness at 20–30%; clouding of consciousness and coma at around 35%; and death at around 50%. However, the outcomes of long-term, low-concentration CO exposures are less well understood. Because of the critical nature of blood flow and $O_2$ delivery to the heart and brain, these organ systems, as well as the lungs (the first organ to come into contact with the pollutant), have received the most attention.

In patients with known coronary artery disease, COHb concentrations as low as 3% exacerbate the development of exercise-induced chest pain. Concentrations as low as 6% are associated with an increase in the number and frequency of premature ventricular contractions during exercise in patients with severe heart disease. Large environmental-exposure cohort studies have confirmed that daily increases in ambient CO concentrations are associated with statistically significant increases in the numbers of hospital admissions for heart disease and congestive heart failure and with increases in deaths from cardiopulmonary illnesses.

Neuropsychiatric (neurological and psychiatric) disorders and cognitive impairments due to long-term, low-concentration CO exposures have been hypothesized in part on the basis of extrapolation from the known acute effects of high-dose CO poisoning and the concomitant subacute and delayed neuropsychological sequelae. In clinical experiments on healthy volunteers, controlled CO exposure was associated with subtle alterations in visual perception when COHb concentrations were above 5%. However the significance of this finding remains unknown. Similar studies have shown measurable but small effects on auditory perception, driving performance, and vigilance.

The role of CO in pulmonary disease is unclear. In the Seattle area, a single-pollutant model showed a 6% increase in the rate of hospital admissions for asthma with each 0.9-ppm increase in CO, but that was concomitant with increases in other air pollutants. In Minneapolis and Toronto, CO concentrations showed only weak and inconsistent associations with total admissions for respiratory diseases.

A fetus is more susceptible to CO than an adult; the $O_2$-hemoglobin dissociation curve is to the left of that in the adult. It is shifted even further to the left by CO exposure. Also, because the half-life of fetal COHb is longer than that of adults, it may take up to five times longer to reduce the concentrations to normal. Other studies have shown that exposure to high concentrations of CO during the last trimester of pregnancy may increase the risk of low birth weights and that exposures to CO and PM during pregnancy may trigger preterm births. A recent study linked CO and $O_3$ exposure during pregnancy to birth defects such as cleft lip and defective heart valves. However, the lead author of the study cautioned that the real culprit may be other pollutants that are dispersed with CO in tailpipe emissions.

Public-health laws are designed to protect the most susceptible people in the population. People with coronary artery disease or other cardiopulmonary diseases, fetuses, infants, and athletes who exercise heavily in high-CO atmospheres are particularly susceptible to experiencing adverse health effects from CO. The evidence summarized above, and described more fully by EPA (2000a), indicates that attainment of the ambient-CO standards can decrease morbidity and mortality from atherosclerotic heart disease. Although less conclusive, there is evidence that attainment of the CO standards will also decrease morbidity from pulmonary disease, neurological disease, fetal loss, and childhood developmental abnormalities. These health benefits translate into economic savings associated with avoided health care and avoided work-time losses as well as intangible savings in life quality.

## CO Exposure

Motor-vehicle emissions are the primary source of CO in outdoor air in populated areas and are associated with the highest outdoor CO exposure in nonsmokers. Outdoor concentrations of CO tend to be higher in urban areas and to increase with the density of vehicles and miles driven. Measurements of ambient CO typically exhibit a bimodal diurnal pattern, with the highest concentrations generally occurring on weekdays during the commuting hours of 7:00–9:00 a.m. and 4:00–6:00 p.m. CO also accumulates in the rider compartments of motor vehicles. Studies have shown that when the concentration near roadways averages 3–4 ppm, the average concentration in the cab is typically 5 ppm.

Most people spend a majority of their time indoors; this is particularly true in Fairbanks and other cold climates during the winter, when ambient CO concentrations tend to be highest. That leads to the question of the relationship between indoor and outdoor concentrations. Air pollution in buildings can come from indoor sources and from air exchange with outdoor ambient pollution. Air exchange may be active, as in the case of a mechanical ventilation system, or passive, as in the case of infiltration associated with temperature or pressure differences between the outside and the interior of a building. CO penetrates freely with infiltration air from the outside and is not removed by building materials or ventilation systems. Furthermore, there are no effective indoor chemical or physical processes for lowering CO on the time scales of interest for exposure and toxic effects.

The relationship between indoor and outdoor CO concentrations can be evaluated with a simple differential mass-balance model that has the following steady-state solution when we combine active ventilation and passive infiltration into a single air-exchange term:

$$C_i = \frac{paC_0}{a+k} + \frac{S}{(a+k)V},$$

where:

$C_i$=indoor concentration, μg/m³;

$C_o$=outdoor concentration, μg/m³;

$p$=penetration coefficient, 0–1;

$a$=air exchange rate, h⁻¹;

$k$=decay rate, h⁻¹;

$S$=mass flux of the indoor source, μg/h; and

$V$=building volume, m³.

Although most areas show a bimodal diurnal pattern with respect to ambient CO concetrations, Fairbanks, Alaska, the focus of this interim report, typically shows a continuous increase throughout the day, with 1-h average CO concentrations peaking at 5:00–6:00 p.m.

For CO, the relationship is simpler because the penetration coefficient (p) is unity and the decay rate (k) is effectively zero. Therefore, the solution is;

$$C_i = C_0 + \frac{S}{aV}.$$

It is clear that in the absence of indoor sources *(S)*, the steady-state indoor concentration of CO will equal the outdoor concentration. When a source of CO is present indoors (for example, from a faulty furnace, an underground parking garage, a kerosene heater, or a smoker), the indoor source adds to the background concentration from the outdoor air. Therefore, buildings do not provide protection from high outdoor concentrations of CO. The idea that buildings provide protection from high outdoor CO concentration is a common misconception.

## Related Pollutants

The incomplete combustion of fossil fuels, which is responsible for CO emissions, also causes emissions of fine particles ($PM_{2.5}$) and toxic organic air contaminants. Epidemiological studies have linked exposure to $PM_{2.5}$ with various adverse health effects, including premature mortality, exacerbation of asthma and other respiratory tract diseases, and decreased lung function. Because of these adverse health effects, EPA issued NAAQS regulating ambient concentrations of $PM_{2.5}$. In addition to the six criteria air pollutants previously regulated, the 1990 amendments of the Clean Air Act (CAAA90) designated 189 toxic air contaminants. Incomplete combustion in mobile sources is estimated to contribute a substantial fraction to the emissions of several toxic air pollutants, including benzene, 1,3-butadiene, and aldehydes. Each of these toxic air pollutants poses some carcinogenic risk. In addition, chronic exposure to benzene is associated with blood disorders; chronic exposure to 1,3-butadiene is associated with cardiovascular disease; and chronic exposure to aldehydes is associated with respiratory problems and eye, nose, and throat irritation. Many emissions-control strategies for CO will also cause reductions in these copollutants and their associated adverse health effects.

In this regard, CO is different from $O_3$, which is highly reactive and therefore rapidly destroyed when infiltrating inside from outdoors.

CO may be a good indicator gas for other pollutants that are emitted at the same time but are not widely measured. The concentrations and spatial distributions of the copollutant species are generally not as well known as for CO. In particular, little data are available about exposure to air toxics present in the ambient environment. CO could be especially useful as an indicator of mobile-source emissions of $PM_{2.5}$ and air toxics, which some studies have shown to be strongly correlated with CO. In one study of emissions from in-use vehicles sampled in Denver, San Antonio, and the Los Angeles area, strong correlations were found between CO and particle emissions ($R^2=0.65$) and between particle and total hydrocarbon (HC) emissions ($R^2=0.78$). The same study demonstrates that emissions of the pollutant species increase with vehicle age and during cold starts. For individual vehicles, however, the correlation among the pollutants is weaker, reflecting the complex mechanisms of formation of the related combustion products.

In another study, three vehicles from Fairbanks were evaluated with the Federal Test Procedure (FTP) for the Urban Dynamometer Driving Schedule. Particle emissions were higher at lower temperatures; the average PM emissions from the three tested automobiles (using regular gasoline) increased from 14.2 to 44.2 mg/mile as the temperature decreased from 20 °F to –20 °F. Most of the particles were released during Phase I of the FTP, representing the first 505 seconds during cold start. The particle emissions also correlated with HC and CO emissions and appeared to be related to rich-operating conditions. Mechanisms of PM formation may include oil consumption.

## CO Emissions from Vehicles

The primary source of CO from vehicles is the incomplete combustion of gasoline in engine cylinders. The fuel-oxidation process (combustion) is the conversion of the fuel to lower-molecular-weight intermediate HCs (including olefins and aromatics) and their conversion to aldehydes and ketones, then to CO, and finally to carbon dioxide ($CO_2$). The initial reactions are faster than the final conversion of CO to $CO_2$. Incomplete conversion of fuel carbon to $CO_2$ results in part from insufficient $O_2$ in the combustion mixture—known as fuel-rich[5] conditions—and insufficient time to oxidize fuel carbon fully to $CO_2$. CO emissions by diesel vehicles are minimal, primarily because of the excess air used in the diesel combustion cycle. Hence, the following discussion is limited to gasoline vehicles.

## Vehicle Technologies

Before the 1980s, carburetors were used to meter fuel in proportion to the intake air of gasoline engines. The design of the carburetor typically ensured.

The ratio of air to fuel mass that provides just enough $O_2$ to convert all the carbon and hydrogen in gasoline to $CO_2$ and water (called stoichiometric) is about 14.7:1. Ratios less than 14.7:1 have more fuel than this optimal mixture and are termed fuel-rich.

Table: National CO Emissions Inventory Estimates for 1999.

| Source Category | Thousands of Short Tons |
| --- | --- |
| Point- or Area-source fuel combustion | 5,322 |
| Electric utilities | 445 |
| Industry | 1,178 |
| Residential wood burning | 3,300 |
| Other | 399 |
| Industrial processes | 7,590 |
| Chemical and allied product manufacturing | 1,081 |
| Metals processing | 1,678 |
| Petroleum and related industries | 366 |
| Waste disposal and recycling | 3,792 |
| Other industrial processes | 599 |
| Onroad vehicles | 49,989 |
| Light-duty gas vehicles and motorcycles | 27,382 |
| Light-duty gas trucks | 16,115 |
| Heavy-duty gas vehicles | 4,262 |
| Diesels | 2,230 |
| Nonroad engines and vehicles | 25,162 |
| Recreational | 3,616 |
| Lawn and garden | 11,116 |
| Aircraft | 1,002 |
| Light commercial | 4,259 |
| Other | 5,169 |
| Miscellaneous | 9,378 |
| Slash or prescribed burning | 6,152 |
| Forest wildfires | 2,638 |
| Other | 588 |
| Total | 97,441 |

That the mixture of air and fuel was adequate to provide vehicle performance, but the carburetor could never be truly optimized for emissions control across the entire operating range of the engine. Carbureted engines featured an automatic choke that increased the fueling rate during cold start and initial operation. Depending on choke calibration, CO and HC emissions could be substantial during cold starts. This type of nonfeedback fuel metering is termed an open-loop system.

Since the middle 1980s, modern computer-controlled engines have used electronic fuel injectors rather than carburetors to deliver fuel to cylinders in automobiles and most light-duty trucks. Using closed-loop control, the engine computer system reads the signal from an $O_2$ sensor in the exhaust system and adjusts the air-to-fuel ratio to help maintain stoichiometric combustion. That feedback provides just enough air to combust the fuel but maintains the maximal catalytic-converter efficiency for control of CO, HC, and nitrogen oxides ($NO_x$).

During hard acceleration and high-speed operations, however, engine computers often use fuel-enrichment strategies to enhance engine performance for short periods and to protect sensitive engine components from high-temperature damage. Likewise, fuel-enrichment strategies are often used during cold starts. Thus, in modern engines, CO emissions are prominent primarily during enrichment associated with heavy loads, hard accelerations, and cold starts.

The onboard diagnostics system installed in model year 1996 and newer vehicles, known as the OBDII system, can help to detect problems during vehicle operations that increase CO emissions. The OBDII system uses sensors to monitor and modify the performance of the engine and emissions-control components. The onboard computer detects signals from the sensors to identify sensor and control-system failures, illuminating the malfunction indicator light on the vehicle dashboard and storing the fault codes (know as diagnostic trouble codes) for later analysis. In a garage setting, mechanics can download the OBDII fault codes from the onboard computer with a diagnostic analyzer ("scan tool"). The codes identify emissions-control systems and components that are malfunctioning. However, some components of OBDII systems (such as exhaust-gas recirculation and $O_2$ sensors) are often disabled by the engine computer under conditions in which the manufacturer cannot guarantee the components' performance. That tends to be the case for vehicles operating at temperatures below 20 °F. It may be prudent to take steps to ensure that manufacturers certify their OBDII systems to lower operating temperatures than currently required. This is especially true if many northern locations begin adopting OBDII inspection and maintenance (I/M) systems, where the OBDII system is relied upon to determine whether a vehicle fails or passes an emissions test. When a significant number of sensors become inoperative, the OBDII system will also have less ability to alert vehicle owners of potential emissions-system failures.

## Cold-Start Emissions

Under cold-start conditions, the engine computer commands the fuel injectors to add excess fuel to the intake air to ensure that enough fuel evaporates to yield a flammable mixture in the engine cylinders. A typical engine-computer strategy injects several times the stoichiometric amount of fuel during the first few engine revolutions, using a fixed fueling schedule to reach idling conditions. Excess fuel continues to be injected until the engine and $O_2$ sensor are warmed up and the exhaust-catalyst inlet temperature reaches about 250–300 °C (482–572 °F), sufficient for the catalyst to oxidize CO to $CO_2$. This open-loop operation, before the catalyst reaches peak efficiency, can continue for several minutes at low ambient temperatures. It is responsible for most of the emissions of CO, air toxics, and unburned HCs from properly operating modern vehicles. Once the engine and emissions-control systems are warmed up, combustion becomes stoichiometric, and CO is converted to $CO_2$ in the catalyst, keeping CO emissions very low under typical operating conditions. Typical warmup times under mild ambient conditions, around 70–80 °F, can be about 1 minute (min) for modern catalysts and even as short as a few seconds for modern close-coupled catalysts (catalysts close to the engine). When ambient temperatures are −20 °F or lower, however, catalyst and engine warmup times can exceed 5 min.

The amount of fuel enrichment required for a cold start of the engine is a function of engine design and ambient air and coolant temperatures. As the temperature in the combustion chamber gets lower, gasoline vapor pressure decreases, and additional fuel is required to ensure ignition. Gasoline sold during the winter in cold climates has a higher vapor pressure than gasoline sold in the summer to assist cold-start ignition. In summer, the vapor pressure is reduced to minimize evaporative emissions.

Below about 0 °F, engine starting becomes more difficult, so external means, such as adjusting the fuel vapor pressure, must be devised to ensure that enough fuel vaporizes. A popular method for assisting engine startup is to use engine-block heaters, which heat the engine coolant to allow easier cold starting. These plug-in heaters reduce the time for engine and catalyst warmup under cold conditions and so help to avoid the extremely high enrichments used by fuel-management control units at low temperatures. Use of plug-in heaters in very cold conditions is also an effective method of reducing CO emissions.

## Strategies to Address Emissions-Control Failures

Although progress has been made in controlling CO emissions from vehicles, some problems remain. One concern is that failures in the operation of emission-control systems typically default to fuel-rich conditions that produce higher CO emissions while allowing engine performance to be maintained. For example, when $O_2$ or temperature sensors are defective, engine computers may default to fuel-rich conditions. Similarly, defective fuel injectors may result in higher CO emissions. A weak spark ignition can cause hard starting, misfires, and poor performance and result in increased CO emissions.

Other common failures in the emissions-control system that affect CO emissions are $O_2$-sensor deterioration, air-injection system defects, and catalyst deterioration. Because engine failures are unavoidable and many such failures cause higher CO emissions, a strategy is needed to identify vehicles with unacceptably high emissions.

Many urban areas use inspection and maintenance (I/M) programs to identify the highest-emitting vehicles and require that they be repaired. High emitters exhibit increased emissions under almost all onroad operating conditions because of failure in emissions-control or fuel-control systems. Not surprisingly, cold-start emissions from high-emitting vehicles are typically also substantially increased. Thus, identification of vehicles as high emitters during the summer months and their repair before winter can yield significant reductions in cold-start emissions. However, some emissions-control system failures may go undetected during an I/M test of a vehicle's tailpipe emissions under idle conditions. Placing a vehicle under a driving load, such as by testing it on a dynamometer, as is done in the IM240 test, has a higher probability of revealing an $O_2$-sensor malfunction and other defects. Remote-sensing systems, which determine CO emissions by measuring absorption of infrared radiation from a beam directed across the roadway, can also be used to identify high emitters.

## Air Quality Models

Air quality modeling is an essential element of air quality management. Models can be used to demonstrate attainment of the NAAQS, evaluate the effects of new construction projects, and conduct further research into what causes pollution episodes and how to predict them. A number of modeling techniques—requiring various levels of scientific expertise, input data, and computing resources—are available for those purposes. The simplest models assume a direct correlation between emissions and ambient pollutant concentrations; the most complicated models resolve temporal and spatial variations in pollutant concentrations and include the effects of meteorology, emissions, chemistry, and topography. Models are also characterized by the size of the problem they address: microscale models simulate pollution from an intersection or point source, mesoscale models simulate metropolitan or multistate pollution, and large-scale models simulate continental or global pollution.

In the attainment demonstration presented in their SIPs, states are required by EPA to model how emissions reductions will lead to the desired air quality improvements. Three types of models have been used to demonstrate attainment of the CO NAAQS: statistical rollback, Gaussian dispersion, and numerical predictive models. The simplest is a statistical rollback model in which the needed reduction in emissions is assumed to be proportional to the required reduction in ambient CO concentrations:

$$\%reduction = \frac{CO_{baseyear} - CO_{NAAQS}}{CO_{baseyear} - CO_{background}},$$

where:

CO$_{baseyear}$ = the second highest 8-h average in the base year;

CO$_{NAAQS}$ = the NAAQS of 9 ppm (or sometimes 9.4 ppm); and

CO$_{background}$ = an average regional background CO in the absence of emissions.

Because no information is needed about meteorology or the spatial distribution of emissions in a nonattainment area, EPA has allowed states to use rollback models to demonstrate attainment in smaller cities, rather than the more resource-intensive dispersion and urban-airshed models described below. Although easy to implement, rollback models do not explicitly consider the role of meteorology or the spatial heterogeneity of CO emissions and concentrations.

A second type of model that has been used for CO-attainment demonstrations is a Gaussian dispersion model, which is typically used to simulate CO concentrations for microscale analysis in the vicinity of intersections or along major traffic corridors. These models simulate how a pollutant is dispersed into the immediately surrounding atmosphere. They assume that the atmospheric concentration of the pollutant is proportional to its emissions and inversely proportional to windspeed and that the resulting spatial distribution of the pollutant is Gaussian. Inputs for dispersion models include meteorological data, such as windspeed and inversion strength in the vicinity of the pollutant source, and temporally resolved emissions. Because predicted concentrations are directly proportional to emissions, the accuracy of emissions measurements is crucial to the modeling process. In the case of modeling of intersections, the emissions inventory may be derived from information about traffic patterns, mean speeds, and vehicle-fleet composition. Larger cities have also used Gaussian dispersion models to evaluate the air quality effects of increasing road capacity or other large construction projects.

Box models, another tool available for microscale analysis of air pollution, have not been used extensively for SIP attainment demonstrations. The "box" is some volume of air into which emissions are injected and in which chemical transformation may take place and air is exchanged with the environment. The air in the box is assumed to be well mixed, so spatial variations in emissions or pollutant concentrations on scales smaller than the box model are not resolved. Box models are particularly useful for understanding how various emissions scenarios and meteorological conditions affect pollutant concentrations. For example, a box model for CO in Anchorage, Alaska, has been used to quantify how mechanical turbulence from roadway traffic might increase the mixing height and reduce CO concentrations on severe-stagnation days. Limitations of the box-model approach include an inability to include spatial variations and a dependence on assumptions to represent meteorological parameters.

The most complicated models used for attainment demonstrations simulate how a pollutant concentration varies with time and space over an entire urban area. These numerical predictive models, generally intended for mesoscale analysis, can simulate emissions from multiple sources and the dispersion, advection, and photochemical reactions of gaseous pollutants in the atmosphere. Numerical predictive models, such as the Urban Airshed Model (UAM), have been used for many years to simulate $O_3$, which is an areawide or mesoscale pollutant. The UAM has also been adapted to simulate CO in Denver, Colorado. Because of the local nature of high-CO episodes, extensive modeling of the entire urban airshed may be unnecessary for CO-attainment demonstrations. Airshed modeling is resource-intensive, requiring detailed knowledge of an area's meteorology (usually based on the output of a mesoscale weather model constrained by observations), spatially and temporally resolved emissions inventories, and measurements of the pollutant at several locations to allow model evaluation. Highly trained personnel are needed to conduct the simulations. However, a simplified approach of this method may be appropriate in some cases.

More complicated models are not always appropriate for attainment demonstrations, but they can be valuable in improving our understanding of the interactions among atmospheric processes. Even better research tools than the numerical predictive models describe above (such as the UAM) are process numerical models, which allow coupling between processes specific to air quality modeling and meteorology. Process numerical models typically are formulated by adding pollutant emissions, chemistry, and transport into an existing meteorological model rather than simply using the meteorological data as a model input. The relatively nonreactive behavior of CO makes it an ideal chemical species for simulation in a weather model. Predictions of CO, for example, can be straightforward in the National Weather Service Eta Model, which has a horizontal grid framework of 12×12 km over the contiguous United States.

Despite advances in air quality modeling capabilities over the last 30 y, many improvements are still possible and needed, particularly in the numerical predictive models, which are used more widely than process numerical models. One problem is that the vertical and horizontal resolution of both types of models is too coarse to capture the variability in pollutant concentrations, which is necessary to identify local hotspots. Most numerical predictive and process numerical models are based on statistical representations of atmospheric motion on scales smaller than the spatial resolution of the models. When unusual meteorological conditions occur, the validity of these representations becomes questionable and could lead to errors in the prediction. Models used for regulatory purposes can suffer the loss of realism as a result of such shortcomings.

Various models have been applied to predict future pollutant concentrations, particularly with the goal of identifying conditions that might create an episode. Numerical predictive models can be used, as can simpler empirical models, which attempt to identify statistically significant relationships between specific air quality variables and a set of predictors. Empirical models typically use regression or neural-network techniques to develop a relationship based on observations of meteorological variables and pollutant concentrations. Future air quality can be predicted by using the output of weather-forecast models as values for the predictors. Meteorological forecasts are disseminated daily by the National Weather Service, but only a few areas in the country provide short-term air quality forecasts, usually only for $O_3$, which is an areawide pollutant.

# Particulates

Suspended particulate matter (SPM) – are microscopic solid or liquid matter suspended in the atmosphere of Earth. The term *aerosol* commonly refers to the particulate/air mixture, as opposed to the particulate matter alone. Sources of particulate matter can be natural or anthropogenic. They have impacts on climate and precipitation that adversely affect human health, in addition to direct inhalation.

Subtypes of atmospheric particles include suspended particulate matter (SPM), thoracic and respirable particles, inhalable coarse particles, which are coarse particles with a diameter between 2.5 and 10 micrometers (µm) ($PM_{10}$), fine particles with a diameter of 2.5 µm or less ($PM_{2.5}$), ultrafine particles, and soot.

The IARC and WHO designate airborne particulates a Group 1 carcinogen. Particulates are the most harmful form of air pollution due to their ability to penetrate deep into the lungs and blood streams unfiltered, causing permanent DNA mutations, heart attacks, respiratory disease, and premature death. In 2013, a study involving 312,944 people in nine European countries revealed that there was no safe level of particulates and that for every increase of 10 µg/m³ in $PM_{10}$, the lung cancer rate rose 22%. The smaller $PM_{2.5}$ were particularly deadly, with a 36% increase in lung cancer per 10 µg/m³ as it can penetrate deeper into the lungs. Worldwide exposure to $PM_{2.5}$ contributed to 4.1 million deaths from heart disease and stroke, lung cancer, chronic lung disease, and respiratory infections in 2016. Overall, ambient particulate matter ranks as the sixth leading risk factor for premature death globally.

## Sources of Atmospheric Particulate Matter

Some particulates occur naturally, originating from volcanoes, dust storms, forest and grassland fires, living vegetation and sea spray. Human activities, such as the burning of fossil fuels in vehicles, stubble burning, power plants, wet cooling towers in cooling systems and various industrial processes, also generate significant amounts of particulates. Coal combustion in developing countries is the primary method for heating homes and supplying energy. Because salt spray over the oceans is the overwhelmingly most common form of particulate in the atmosphere, *anthropogenic* aerosols—those made by human activities—currently account for about 10 percent of the total mass of aerosols in our atmosphere.

## Composition

The composition of aerosols and particles depends on their source. Wind-blown mineral dust tends to be made of mineral oxides and other material blown from the Earth's crust; this particulate is light-absorbing. Sea salt is considered the second-largest contributor in the global aerosol budget, and consists mainly of sodium chloride originated from sea spray; other constituents of atmospheric sea salt reflect the composition of sea water, and thus include magnesium, sulfate, calcium, potassium, etc. In addition, sea spray aerosols may contain organic compounds, which influence their chemistry. The drift/mist emissions from the wet cooling towers is also source of particulate matter as they are widely used in industry and other sectors for dissipating heat in cooling systems.

Secondary particles derive from the oxidation of primary gases such as sulfur and nitrogen oxides into sulfuric acid (liquid) and nitric acid (gaseous). The precursors for these aerosols—i.e. the gases from which they originate—may have an anthropogenic origin (from fossil fuel or coal combustion) and a natural biogenic origin. In the presence of ammonia, secondary aerosols often take the form of ammonium salts; i.e. ammonium sulfate and ammonium nitrate (both can be dry or in aqueous solution); in the absence of ammonia, secondary compounds take an acidic form as sulfuric acid (liquid aerosol droplets) and nitric acid (atmospheric gas), all of which may contribute to the health effects of particulates.

Secondary sulfate and nitrate aerosols are strong light-scatterers. This is mainly because the presence of sulfate and nitrate causes the aerosols to increase to a size that scatters light effectively.

Organic matter (OM) can be either primary or secondary, the latter part deriving from the oxidation of VOCs; organic material in the atmosphere may either be biogenic or anthropogenic. Organic matter influences the atmospheric radiation field by both scattering and absorption. Another important aerosol type is elemental carbon (EC, also known as *black carbon*, BC): this aerosol type includes strongly light-absorbing material and is thought to yield large positive radiative forcing. Organic matter and elemental carbon together constitute the carbonaceous fraction of aerosols. Secondary organic aerosols (SOAs), tiny "tar balls" resulting from combustion products of internal combustion engines, have been identified as a danger to health.

The chemical composition of the aerosol directly affects how it interacts with solar radiation. The chemical constituents within the aerosol change the overall refractive index. The refractive index will determine how much light is scattered and absorbed.

The composition of particulate matter that generally causes visual effects such as smog consists of sulfur dioxide, nitrogen oxides, carbon monoxide, mineral dust, organic matter, and elemental carbon also known as black carbon or soot. The particles are hygroscopic due to the presence of sulfur, and $SO_2$ is converted to sulfate when high humidity and low temperatures are present. This causes the reduced visibility and yellow color.

## Size Distribution of Particulates

Aerosol particles of natural origin (such as windblown dust) tend to have a larger radius than human-produced aerosols such as particle pollution. The false-color maps image show where there are natural aerosols, human pollution, or a mixture of both, monthly.

Among the most obvious patterns that the size distribution time series shows is that in the planet's most southerly latitudes, nearly all the aerosols are large, but in the high northern latitudes, smaller aerosols are very abundant. Most of the Southern Hemisphere is covered by ocean, where the largest source of aerosols is natural sea salt from dried sea spray. Because the land is concentrated in the Northern Hemisphere, the amount of small aerosols from fires and human activities is greater there than in the Southern Hemisphere. Over land, patches of large-radius aerosols appear over deserts and arid regions, most prominently, the Sahara Desert in North Africa and the Arabian Peninsula, where dust storms are common. Places where human-triggered or natural fire activity is common (land-clearing fires in the Amazon from August–October, for example, or lightning-triggered fires in the forests of northern Canada in Northern Hemisphere summer) are dominated by smaller aerosols. Human-produced (fossil fuel) pollution is largely responsible for the areas of small aerosols over developed areas such as the eastern United States and Europe, especially in their summer.

Satellite measurements of aerosols, called aerosol optical thickness, are based on the fact that the particles change the way the atmosphere reflects and absorbs visible and infrared light. As shown in the image an optical thickness of less than 0.1 (palest yellow) indicates a crystal clear sky with maximum visibility, whereas a value of 1 (reddish brown) indicates very hazy conditions.

## Deposition Processes

In general, the smaller and lighter a particle is, the longer it will stay in the air. Larger particles (greater than 10 micrometers in diameter) tend to settle to the ground by gravity in a matter of hours whereas the smallest particles (less than 1 micrometer) can stay in the atmosphere for weeks and are mostly removed by precipitation. Diesel particulate matter is highest near the source of emission. Any info regarding DPM and the atmosphere, flora, height, and distance from major sources would be useful to determine health effects.

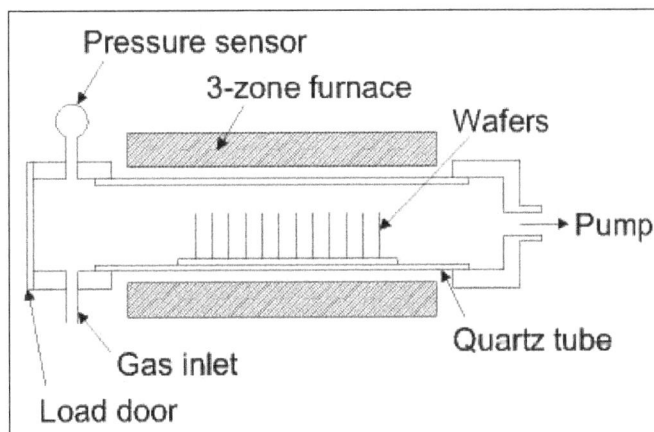

Deposition Processes.

## Controlling Technologies

A complicated blend of solid and liquid particles result in particulate matter and these particulate matter emissions are highly regulated in most industrialized countries. Due to environmental concerns, most industries are required to operate some kind of dust collection system to control particulate emissions. These systems include inertial collectors (cyclonic separators), fabric filter collectors (baghouses), wet scrubbers, and electrostatic precipitators.

Cyclonic separators are useful for removing large, coarse particles and are often employed as a first step or "pre-cleaner" to other more efficient collectors. Well-designed cyclonic separators can be very efficient in removing even fine particulates, and may be operated continuously without requiring frequent shutdowns for maintenance.

Fabric filters or baghouses are the most commonly employed in general industry. They work by forcing dust laden air through a bag shaped fabric filter leaving the particulate to collect on the outer surface of the bag and allowing the now clean air to pass through to either be exhausted into the atmosphere or in some cases recirculated into the facility. Common fabrics include polyester and fiberglass and common fabric coatings include PTFE (commonly known as Teflon). The excess dust buildup is then cleaned from the bags and removed from the collector.

Wet scrubbers pass the dirty air through a scrubbing solution (usually a mixture of water and other compounds) allowing the particulate to attach to the liquid molecules. Electrostatic precipitators electrically charge the dirty air as it passes through. The now charged air then passes through large electrostatic plates which attract the charged particle in the airstream collecting them and leaving the now clean air to be exhausted or recirculated. Besides removing particulates from the source of the pollution, it can also be cleaned in the open air.

## Climate Effects

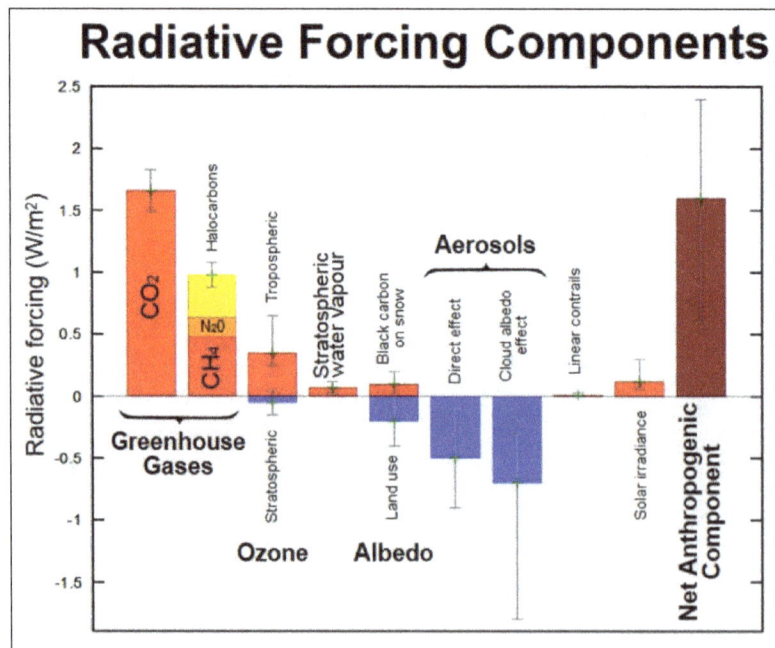

2005 Radiative forcings and uncertainties as estimated by the IPCC.

Atmospheric aerosols affect the climate of the earth by changing the amount of incoming solar radiation and outgoing terrestrial longwave radiation retained in the earth's system. This occurs through several distinct mechanisms which are split into direct, indirect and semi-direct aerosol effects. The aerosol climate effects are the biggest source of uncertainty in future climate predictions. While the radiative forcing due to greenhouse gases may be determined to a reasonably high degree of accuracy the uncertainties relating to aerosol radiative forcings remain large, and rely to a large extent on the estimates from global modelling studies that are difficult to verify at the present time.

## Aerosol Radiative Effects

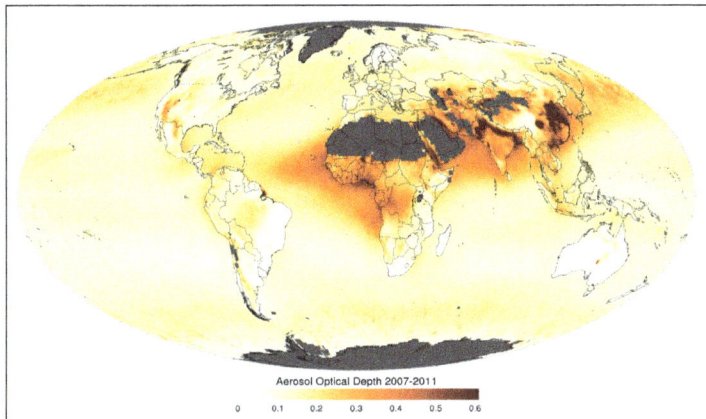

These maps show average monthly aerosol amounts around the world based on observations from the Moderate Resolution Imaging Spectroradiometer (MODIS) on NASA's Terra satellite.

## Direct Effect

Particulates in the air causing shades of grey and pink in Mumbai during sunset.

The direct aerosol effect consists of any direct interaction of radiation with atmospheric aerosols, such as absorption or scattering. It affects both short and longwave radiation to produce a net negative radiative forcing. The magnitude of the resultant radiative forcing due to the direct effect of an aerosol is dependent on the albedo of the underlying surface, as this affects the net amount of radiation absorbed or scattered to space. e.g. if a highly scattering aerosol is above a surface of low albedo it has a greater radiative forcing than if it was above a surface of high albedo. The converse is true of absorbing aerosol, with the greatest radiative forcing arising from a highly absorbing aerosol over a surface of high albedo. The direct aerosol effect is a first order effect and is therefore classified as a radiative forcing by the IPCC. The interaction of an aerosol with radiation is quantified by the single-scattering albedo (SSA), the ratio of scattering alone to scattering plus absorption (*extinction*) of radiation by a particle. The SSA tends to unity if scattering dominates, with relatively little absorption, and decreases as absorption increases, becoming zero for infinite absorption. For example, the sea-salt aerosol has an SSA of 1, as a sea-salt particle only scatters, whereas soot has an SSA of 0.23, showing that it is a major atmospheric aerosol absorber.

## Indirect Effect

The Indirect aerosol effect consists of any change to the earth's radiative budget due to the modification of clouds by atmospheric aerosols, and consists of several distinct effects. Cloud droplets form onto pre-existing aerosol particles, known as cloud condensation nuclei (CCN).

For any given meteorological conditions, an increase in CCN leads to an increase in the number of cloud droplets. This leads to more scattering of shortwave radiation i.e. an increase in the albedo of the cloud, known as the Cloud albedo effect, First indirect effect or Twomey effect. Evidence supporting the cloud albedo effect has been observed from the effects of ship exhaust plumes and biomass burning on cloud albedo compared to ambient clouds. The Cloud albedo aerosol effect is a first order effect and therefore classified as a radiative forcing by the IPCC.

An increase in cloud droplet number due to the introduction of aerosol acts to reduce the cloud droplet size, as the same amount of water is divided into more droplets. This has the effect of suppressing precipitation, increasing the cloud lifetime, known as the cloud lifetime aerosol effect, second indirect effect or Albrecht effect. This has been observed as the suppression of drizzle in ship exhaust plume compared to ambient clouds, and inhibited precipitation in biomass burning plumes. This cloud lifetime effect is classified as a climate feedback (rather than a radiative forcing) by the IPCC due to the interdependence between it and the hydrological cycle. However, it has previously been classified as a negative radiative forcing.

## Semi-direct Effect

The Semi-direct effect concerns any radiative effect caused by absorbing atmospheric aerosol such as soot, apart from direct scattering and absorption, which is classified as the direct effect. It encompasses many individual mechanisms, and in general is more poorly defined and understood than the direct and indirect aerosol effects. For instance, if absorbing aerosols are present in a layer aloft in the atmosphere, they can heat surrounding air which inhibits the condensation of water vapour, resulting in less cloud formation. Additionally, heating a layer of the atmosphere relative to the surface results in a more stable atmosphere due to the inhibition of atmospheric convection. This inhibits the convective uplift of moisture, which in turn reduces cloud formation. The heating of the atmosphere aloft also leads to a cooling of the surface, resulting in less evaporation of surface water. The effects described here all lead to a reduction in cloud cover i.e. an increase in planetary albedo. The semi-direct effect classified as a climate feedback) by the IPCC due to the interdependence between it and the hydrological cycle. However, it has previously been classified as a negative radiative forcing.

## Roles of different Aerosol Species

## Sulfate Aerosol

Sulfate aerosol has two main effects, direct and indirect. The direct effect, via albedo, is a cooling effect that slows the overall rate of global warming: the IPCC's best estimate of the radiative forcing is −0.4 watts per square meter with a range of −0.2 to −0.8 W/m² but there are substantial uncertainties. The effect varies strongly geographically, with most cooling believed to be at and downwind of major industrial centres. Modern climate models addressing the attribution of recent climate change take into account sulfate forcing, which appears to account (at least partly) for the slight drop in global temperature in the middle of the 20th century. The indirect effect (via the aerosol acting as cloud condensation nuclei, CCN, and thereby modifying the cloud properties -albedo and lifetime-) is more uncertain but is believed to be a cooling.

## Black Carbon

Black carbon (BC), or carbon black, or elemental carbon (EC), often called soot, is composed of pure carbon clusters, skeleton balls and buckyballs, and is one of the most important absorbing aerosol species in the atmosphere. It should be distinguished from organic carbon (OC): clustered or aggregated organic molecules on their own or permeating an EC buckyball. BC from fossil fuels is estimated by the IPCC in the Fourth Assessment Report of the IPCC, 4AR, to contribute a global mean radiative forcing of +0.2 W/m² (was +0.1 W/m² in the Second Assessment Report of the IPCC, SAR), with a range +0.1 to +0.4 W/m². Bond et al., however, states that "the best estimate for the industrial-era (1750 to 2005) direct radiative forcing of atmospheric black carbon is +0.71 W/m² with 90% uncertainty bounds of (+0.08, +1.27) W/m²" with "total direct forcing by all black carbon sources, without subtracting the preindustrial background, is estimated as +0.88 (+0.17, +1.48) W/m²".

## Instances of Aerosol affecting Climate

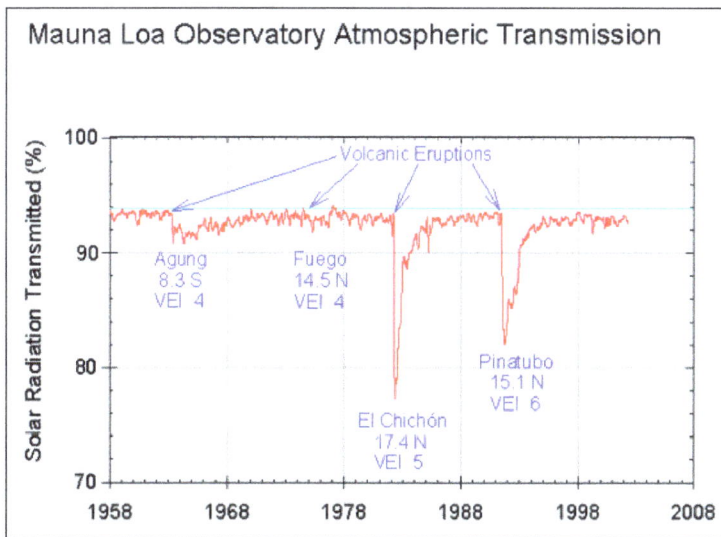

Solar radiation reduction due to volcanic eruptions.

Volcanoes are a large natural source of aerosol and have been linked to changes in the earth's climate often with consequences for the human population. Eruptions linked to changes in climate include the 1600 eruption of Huaynaputina which was linked to the Russian famine of 1601 - 1603, leading to the deaths of two million, and the 1991 eruption of Mount Pinatubo which caused a global cooling of approximately 0.5 °C lasting several years. Research tracking the effect of light-scattering aerosols in the stratosphere during 2000 and 2010 and comparing its pattern to volcanic activity show a close correlation. Simulations of the effect of anthropogenic particles showed little influence at present levels.

Aerosols are also thought to affect weather and climate on a regional scale. The failure of the Indian Monsoon has been linked to the suppression of evaporation of water from the Indian Ocean due to the semi-direct effect of anthropogenic aerosol.

Recent studies of the Sahel drought and major increases since 1967 in rainfall over the Northern Territory, Kimberley, Pilbara and around the Nullarbor Plain have led some scientists to conclude that the aerosol haze over South and East Asia has been steadily shifting tropical rainfall in both hemispheres southward.

The latest studies of severe rainfall decline over southern Australia since 1997 have led climatologists there to consider the possibility that these Asian aerosols have shifted not only tropical but also midlatitude systems southward.

## Health Effects

Air pollution measurement station in Emden.

## Size, Shape and Solubility Matter

The size of the particle is a main determinant of where in the respiratory tract the particle will come to rest when inhaled. Larger particles are generally filtered in the nose and throat via cilia and mucus, but particulate matter smaller than about 10 micrometers, can settle in the bronchi and lungs and cause health problems. The 10-micrometer size does not represent a strict boundary between respirable and non-respirable particles, but has been agreed upon for monitoring of airborne particulate matter by most regulatory agencies. Because of their small size, particles on the order of 10 micrometers or less (coarse particulate matter, $PM_{10}$) can penetrate the deepest part of the lungs such as the bronchioles or alveoli; when asthmatics are exposed to these conditions it can trigger bronchoconstriction.

Similarly, so called fine particulate matter ($PM_{2.5}$), tend to penetrate into the gas exchange regions of the lung (alveolus), and very small particles (ultrafine particulate matter, $PM_{0.1}$) may pass through the lungs to affect other organs. Penetration of particles is not wholly dependent on their size; shape and chemical composition also play a part. To avoid this complication, simple nomenclature is used to indicate the different degrees of relative penetration of a PM particle into the cardiovascular system. Inhalable particles penetrate no further than the bronchi as they are filtered out by the cilia. Thoracic particles can penetrate right into terminal bronchioles whereas $PM_{0.1}$, which can penetrate to alveoli, the gas exchange area, and hence the circulatory system are termed respirable particles. In analogy, the inhalable dust fraction is the fraction of dust entering nose and mouth which may be deposited anywhere in the respiratory tract. The thoracic fraction is the fraction that enters the thorax and is deposited within the lung's airways. The respirable fraction is what is deposited in the gas exchange regions (alveoli).

The smallest particles, less than 100 nanometers (nanoparticles), may be even more damaging to the cardiovascular system. Nanoparticles can pass through cell membranes and migrate into other organs, including the brain. Particles emitted from modern diesel engines (commonly referred to as Diesel Particulate Matter, or DPM) are typically in the size range of 100 nanometers (0.1 micrometer). These soot particles also carry carcinogens like benzopyrenes adsorbed on their surface. Particulate mass is not a proper measure of the health hazard, because one particle of 10 μm diameter has approximately the same mass as 1 million particles of 100 nm diameter, but is much less hazardous, as it is unlikely to enter the alveoli. Legislative limits for engine emissions based on mass are therefore not protective. Proposals for new regulations exist in some countries, with suggestions to limit the particle *surface area* or the *particle count* (numerical quantity) instead.

The site and extent of absorption of inhaled gases and vapors are determined by their solubility in water. Absorption is also dependent upon air flow rates and the partial pressure of the gases in the inspired air. The fate of a specific contaminant is dependent upon the form in which it exists (aerosol or particulate). Inhalation also depends upon the breathing rate of the subject.

Another complexity not entirely documented is how the shape of PM can affect health, except for the needle-like shape of asbestos which can lodge itself in the lungs. Geometrically angular shapes have more surface area than rounder shapes, which in turn affects the binding capacity of the particle to other, possibly more dangerous substances.

## Health Problems

The effects of inhaling particulate matter that has been widely studied in humans and animals include asthma, lung cancer, respiratory diseases, cardiovascular disease, premature delivery, birth defects, low birth weight, and premature death.

Inhalation of $PM_{2.5}$ – $PM_{10}$ is associated with elevated risk of adverse pregnancy outcomes, such as low birth weight. Maternal $PM_{2.5}$ exposure during pregnancy is also associated with high blood pressure in children. Exposure to $PM_{2.5}$ has been associated with greater reductions in birth weight than exposure to $PM_{10}$. PM exposure can cause inflammation, oxidative stress, endocrine disruption, and impaired oxygen transport access to the placenta, all of which are mechanisms for heightening the risk of low birth weight. Overall epidemiologic and toxicological evidence suggests that a causal relationship exists between long-term exposures to $PM_{2.5}$ and developmental outcomes (i.e. low birth weight). However, studies investigating the significance of trimester-specific exposure have proven to be inconclusive, and results of international studies have been inconsistent in drawing associations of prenatal particulate matter exposure and low birth weight. As perinatal outcomes have been associated with lifelong health and exposure to particulate matter is widespread, this issue is of critical public health importance and additional research will be essential to inform public policy on the matter.

Increased levels of fine particles in the air as a result of *anthropogenic* particulate air pollution "is consistently and independently related to the most serious effects, including lung cancer and other cardiopulmonary mortality." A large number of deaths and other health problems associated with particulate pollution was first demonstrated in the early 1970s and has been reproduced many times since. PM pollution is estimated to cause 22,000–52,000 deaths per year in the United States contributed to ~370,000 premature deaths in Europe during 2005. and 3.22 million deaths globally in 2010 per the global burden of disease collaboration.

A 2002 study indicated that $PM_{2.5}$ leads to high plaque deposits in arteries, causing vascular inflammation and atherosclerosis – a hardening of the arteries that reduces elasticity, which can lead to heart attacks and other cardiovascular problems. A 2014 meta analysis reported that long term exposure to particulate matter is linked to coronary events. The study included 11 cohorts participating in the European Study of Cohorts for Air Pollution Effects (ESCAPE) with 100,166 participants, followed for an average of 11.5 years. An increase in estimated annual exposure to PM 2.5 of just 5 µg/m³ was linked with a 13% increased risk of heart attacks. In 2017 a study revealed that PM not only affects human cells and tissues, but also impacts bacteria which cause disease in humans. This study concluded that biofilm formation, antibiotic tolerance, and colonisation of both Staphylococcus aureus and Streptococcus pneumoniae was altered by Black Carbon exposure.

The World Health Organization (WHO) estimated in 2005 that "fine particulate air pollution (PM(2.5)), causes about 3% of mortality from cardiopulmonary disease, about 5% of mortality from cancer of the trachea, bronchus, and lung, and about 1% of mortality from acute respiratory infections in children under 5 years, worldwide". A 2011 study concluded that traffic exhaust is the single most serious preventable cause of heart attack in the general public, the cause of 7.4% of all attacks.

The largest US study on acute health effects of coarse particle pollution between 2.5 and 10 micrometers in diameter was published 2008 and found an association with hospital admissions for cardiovascular diseases but no evidence of an association with the number of hospital admissions for respiratory diseases. After taking into account fine particle levels ($PM_{2.5}$ and less), the association with coarse particles remained but was no longer statistically significant, which means the effect is due to the subsection of fine particles.

Particulate matter studies in Bangkok Thailand from 2008 indicated a 1.9% increased risk of dying from cardiovascular disease, and 1.0% risk of all disease for every 10 micrograms per cubic meter. Levels averaged 65 in 1996, 68 in 2002, and 52 in 2004. Decreasing levels may be attributed to conversions of diesel to natural gas combustion as well as improved regulations.

The Mongolian government agency recorded a 45% increase in the rate of respiratory illness in the past five years. Bronchial asthma, chronic obstructive pulmonary disease and interstitial pneumonia were the most common ailments treated by area hospitals. Levels of premature death, chronic bronchitis, and cardiovascular disease are increasing at a rapid rate.

A study in 2000 conducted in the U.S. explored how fine particulate matter may be more harmful than coarse particulate matter. The study was based on six different cities. They found that deaths and hospital visits that were caused by particulate matter in the air were primarily due fine particulate matter.

### Effects on Vegetation

Particulate matter can clog stomatal openings of plants and interfere with photosynthesis functions. In this manner, high particulate matter concentrations in the atmosphere can lead to growth stunting or mortality in some plant species.

# Ammonia

Ammonia ($NH_3$) is a highly reactive and soluble alkaline gas. It originates from both natural and anthropogenic sources, with the main source being agriculture, e.g. manures, slurries and fertiliser application.

Excess nitrogen can cause eutrophication and acidification effects on semi-natural ecosystems, which in turn can lead to species composition changes and other deleterious effects.

Ammonia comes from the breakdown and volatilisation of urea. Emissions and deposition vary spatially, with "emission hot-spots" associated with high-density intensive farming practices. Other agriculture-related emissions of ammonia include biomass burning or fertiliser manufacture. Ammonia is also emitted from a range of non-agricultural sources, such as catalytic converters in petrol cars, landfill sites, sewage works, composting of organic materials, combustion, industry and wild mammals and birds.

### Emissions

At the turn of the 21st century, total ammonia emissions in the UK were estimated to be 283 kt N yr-1 with 228 kt coming from agricultural sources. In 2010 the agricultural sector was responsible for 89% of UK $NH_3$ emissions. National $NH_3$ emissions in the UK are mapped at a 5 km grid resolution, using the AENEID model for agricultural sources, and at a 1 km or 5 km grid resolution for non-agricultural sources.

Emissions trends have mostly been downward since peak in late 1980s and early 1990s but have now flattened. As the climate warms, volatilisation of ammonia emissions will lead to a further rise in ammonia concentrations.

High emission areas with intensive dairy farming can be distinguished from low emission areas with extensive sheep and beef farming or "hot-spot" patterns associated with intensive pig and poultry farming. Emissions from agricultural sources vary temporally with agricultural practice. Seasonal variation is also associated with climate; volatilisation being highest when it is warmer. Some non-agricultural emission sources (e.g. seabird colonies) contribute only small amounts to the overall $NH_3$ emissions in the UK but, due to their location, are often the dominant emission source in remote and otherwise "clean" areas. Larger seabird colonies have been shown to emit similar amounts of $NH_3$ to large intensive poultry farms.

## Atmospheric Interactions

Atmospheric ammonia has impacts on both local and international (transboundary) scales. In the atmosphere ammonia reacts with acid pollutants such as the products of $SO_2$ and $NO_x$ emissions to produce fine ammonium ($NH_4^+$) containing aerosol. While the lifetime of $NH_3$ is relatively short (<10-100 km), $NH_4^+$ may be transferred much longer distances (100->1000 km). Hence $NH_3$ emissions contribute to international transboundary air pollutant issues addressed by the UNECE Convention on Long Range Transboundary Pollution.

In addition to the transboundary effects, $NH_3$ has substantial impacts at a local level: emissions occur at ground level in the rural environment and $NH_3$ is rapidly deposited. As a result some of the most acute problems of $NH_3$ deposition are for small relict nature reserves located in intensive agricultural landscapes.

Ammonia can be volatilised, emitted into the atmosphere when the surface concentration exceeds that of the surrounding air. Losses of $NH_3$ by volatilisation from the application of nitrogen (N) fertilisers range from negligible amounts to >50% of the applied fertiliser N, depending on fertiliser/manure type (e.g. urea higher volatilization rates than ammonium nitrate), application practice (e.g. injection, surface application) and environmental conditions. Solubility and dissolution processes primarily drive the magnitude of $NH_3$ emissions, higher in warm drying conditions and smaller in cool wet conditions.

## Concentrations and Deposition

Ammonia concentrations are monitored, and show large spatial variability, reflecting a combination of the large number of ground level sources, primarily related to livestock farming, and the very reactive nature of gaseous $NH_3$. Concentrations of $NH_3$ range from 10 μg m-3 in areas of intensive livestock production, especially dairy and beef production, to 0.1 μg m-3 in the Scottish Highlands, especially in the north-west of Scotland and in the Hebrides.

These concentrations can be used to estimate deposition although deposition varies with ecosystem type and meteorology. Due to the varying affinity and compensation points of ammonia for different habitats, expressed in differences in mean deposition velocities, the rates of ammonia deposition vary greatly between habitat types.

Maps of concentrations and depositions across the UK are mapped using the FRAME model and calibrated using the measured $NH_3$ values at monitoring stations. This means that maps of $NH_3$ dry deposition need to be interpreted with care, noting whether they refer to inputs to specific habitat types (e.g. woodland, shrublands and croplands) or net dry deposition averaged over entire grid squares. For the purpose of assessing critical loads exceedance, deposition values for the relevant habitats need to be used, rather than grid averages.

Areas at risk from ammonia/nitrogen impacts include those close to point sources and areas within intensive agricultural regions which see elevated ammonia concentrations.

## Effects

Effects of ammonia have been established from transect studies downwind of significant $NH_3$ sources and a field release. Ammonia can be taken up through the leaves via stomata, increasing the potential for nutrient N uptake. The consequences of foliar uptake and processing of an alkaline gas for cellular functions, appear to drive the deleterious effects of $NH_3$ on terrestrial plants. Alkalinity is also thought to be a key driver for $NH_3$ effects on epiphytic lichens. Atmospheric $NH_3$ also impacts as $NH_4^+$, when the $NH_3$ deposits to plant surfaces, dissolves and is washed into the soil where it can increase soil acidity and interfere with base cation uptake. Effects represent the combined effects of uptake through shoots as $NH_3/NH_4^+$ and roots as $NH_4^+$.

Negative effects on vegetation occur via direct toxicity, when uptake exceeds detoxification capacity and, via N accumulation, which increases the likelihood of detrimental interactions with other abiotic and biotic stressors. Ammonia can also enrich a system with nitrogen putting under-storey species at risk as they become shaded by the expansion of nitrophiles (N loving plants) that use the additional N to increase productivity and expand the over-storey. Nitrogen enrichment affects competition for resources, favouring fast growing, tall species with rapid N assimilation rates. Mosses and lichens are most at risk, they have limited detoxification capacity relative to their uptake potential and a large surface area relative to mass.

Many lichen species are sensitive to even small increases in $NH_3$ concentrations above c. 1µg m$^{-3}$. Current evidence suggests that the absence of acidophytic lichens (lichens loving acid conditions) from twigs and trunks of acid-barked trees, growing in $NH_3$ rich environments, is due to $NH_3$ neutralizing the bark pH. Sheppard et al. found that monthly $NH_3$ concentrations > 20 µg m$^{-3}$ decimated Cladonia portentosa populations in less than one year and that after three years the concentration had fallen to < 3 µg m$^{-3}$. Wet deposited $NH_4^+$ caused only restricted damage.

In mosses, $NH_3$ exposure can increase both the N and amino acid content of ectohydric pleurocarpous mosses. Elevations in N and amino acid content have been proposed as a well coupled indicator of NH-N deposition. Moss species differ with respect to their N uptake, and presumably their tolerance. Some Sphagnum (bog mosses) appear to be very sensitive, especially those that lack the red-orange pigments, carotenoids, that protect against oxidative stress. Overall dry deposited ammonia-N drives species composition change and reduces species cover and diversity, much faster than the same unit of N in wet deposition.

Attributing both specific effects in the field and indicators can be challenging because ammonia is a form of nitrogen which is an essential plant growth nutrient. In addition, some of the effects are difficult to separate from those caused by management, or lack of shading of the under-storey.

Effects on vegetation are:

- Eutrophication leading to changes in species assemblages; increase in N loving species (e.g. grasses) and species that can up regulate their carbon assimilation at the expense of species that are conservative in their N use.

- Shift in dominance from mosses, lichens and ericoids (heath species) towards grasses like Deschampsia flexuosa, Molinia caerulea and ruderal species, e.g. Chamerion angustifolium, Rumex acetosella, Rubus idaeus.

- Increased risk of frost damage in spring.

- Increased winter desiccation levels in Calluna and summer drought stress.

- Increase in N loving epiphytes, e.g. Xanthoria parietina, at the expense of epiphytes that prefer acid bark.

- Increased incidence of pest and pathogen attack, e.g. heather beetle outbreaks.

- Direct damage and death of sensitive species, e.g. lichens and mosses, Sphagnum, Pleurozium schreberi.

- Reduced root growth and mycorrhizal infection leading to reduced nutrient uptake, sensitivity to drought and nutrient imbalance with respect to N that is taken up via the foliage.

- Increase in soil pH follows acidification.

- Ammonia excess will lead to increases in nitrification and denitrification, contributing to greenhouse gas emissions.

# Volatile Organic Compound

Volatile organic compounds (VOCs) are emitted as gases from certain solids or liquids. VOCs include a variety of chemicals, some of which may have short- and long-term adverse health effects. Concentrations of many VOCs are consistently higher indoors (up to ten times higher) than outdoors. VOCs are emitted by a wide array of products numbering in the thousands.

Organic chemicals are widely used as ingredients in household products. Paints, varnishes and wax all contain organic solvents, as do many cleaning, disinfecting, cosmetic, degreasing and hobby products. Fuels are made up of organic chemicals. All of these products can release organic compounds while you are using them, and, to some degree, when they are stored.

## Sources of VOCs

Household products, including:

- Paints, paint strippers and other solvents.

- Wood preservatives.

- Aerosol sprays.

- Cleansers and disinfectants.

- Moth repellents and air fresheners.

- Stored fuels and automotive products.

- Hobby supplies.

- Dry-cleaned clothing.

- Pesticide.

Other products, including:

- Building materials and furnishings.

- Office equipment such as copiers and printers, correction fluids and carbonless copy paper.

- Graphics and craft materials including glues and adhesives, permanent markers and pho-tographic solutions.

## Health Effects

Health effects may include:

- Eye, nose and throat irritation.

- Headaches, loss of coordination and nausea.

- Damage to liver, kidney and central nervous system.

- Some organics can cause cancer in animals, some are suspected or known to cause cancer in humans.

Key signs or symptoms associated with exposure to VOCs include:

- Conjunctival irritation.

- Nose and throat discomfort.

- Headache.

- Allergic skin reaction.

- Dyspnea.

- Declines in serum cholinesterase levels.

- Nausea.

- Emesis.

- Epistaxis.

- Fatigue.

- Dizziness.

The ability of organic chemicals to cause health effects varies greatly from those that are highly toxic, to those with no known health effect.

As with other pollutants, the extent and nature of the health effect will depend on many factors including level of exposure and length of time exposed. Among the immediate symptoms that some people have experienced soon after exposure to some organics include:

- Eye and respiratory tract irritation.

- Headaches.

- Dizziness.

- Visual disorders and memory impairment.

At present, not much is known about what health effects occur from the levels of organics usually found in homes.

## Levels in Homes

Studies have found that levels of several organics average 2 to 5 times higher indoors than outdoors. During and for several hours immediately after certain activities, such as paint stripping, levels may be 1,000 times background outdoor levels.

## Steps to Reduce Exposure

- Increase ventilation when using products that emit VOCs.

- Meet or exceed any label precautions.

- Do not store opened containers of unused paints and similar materials within the school.

- Formaldehyde, one of the best known VOCs, is one of the few indoor air pollutants that can be readily measured.

    o Identify, and if possible, remove the source.

    o If not possible to remove, reduce exposure by using a sealant on all exposed surfaces of paneling and other furnishings.

- Use integrated pest management techniques to reduce the need for pesticides.

- Use household products according to manufacturer's directions.

- Make sure you provide plenty of fresh air when using these products.

- Throw away unused or little-used containers safely; buy in quantities that you will use soon.

- Keep out of reach of children and pets.

- Never mix household care products unless directed on the label.

Follow label instructions carefully: Potentially hazardous products often have warnings aimed at reducing exposure of the user. For example, if a label says to use the product in a well-ventilated area, go outdoors or in areas equipped with an exhaust fan to use it. Otherwise, open up windows to provide the maximum amount of outdoor air possible.

Throw away partially full containers of old or unneeded chemicals safely: Because gases can leak even from closed containers, this single step could help lower concentrations of organic chemicals in your home. (Be sure that materials you decide to keep are stored not only in a well-ventilated area but are also safely out of reach of children.) Do not simply toss these unwanted products in the garbage can. Find out if your local government or any organization in your community sponsors special days for the collection of toxic household wastes. If such days are available, use them to dispose of the unwanted containers safely. If no such collection days are available, think about organizing one.

Buy limited quantities: If you use products only occasionally or seasonally, such as paints, paint strippers and kerosene for space heaters or gasoline for lawn mowers, buy only as much as you will use right away.

Keep exposure to emissions from products containing methylene chloride to a minimum: Consumer products that contain methylene chloride include paint strippers, adhesive removers and aerosol spray paints. Methylene chloride is known to cause cancer in animals. Also, methylene chloride is converted to carbon monoxide in the body and can cause symptoms associated with exposure to carbon monoxide. Carefully read the labels containing health hazard information and cautions on the proper use of these products. Use products that contain methylene chloride outdoors when possible; use indoors only if the area is well ventilated.

Keep exposure to benzene to a minimum: Benzene is a known human carcinogen. The main indoor sources of this chemical are:

- Environmental tobacco smoke.

- Stored fuels.

- Paint supplies.

- Automobile emissions in attached garages.

Actions that will reduce benzene exposure include:

- Eliminating smoking within the home.

- Providing for maximum ventilation during painting.

- Discarding paint supplies and special fuels that will not be used immediately.

Keep exposure to perchloroethylene emissions from newly dry-cleaned materials to a minimum: Perchloroethylene is the chemical most widely used in dry cleaning. In laboratory studies, it has been shown to cause cancer in animals. Recent studies indicate that people breathe low levels of this chemical both in homes where dry-cleaned goods are stored and as they wear dry-cleaned clothing. Dry cleaners recapture the perchloroethylene during the dry-cleaning process so they can save money by re-using it, and they remove more of the chemical during the pressing and finishing processes. Some dry cleaners, however, do not remove as much perchloroethylene as possible all of the time.

Taking steps to minimize your exposure to this chemical is prudent:

- If dry-cleaned goods have a strong chemical odor when you pick them up, do not accept them until they have been properly dried.

- If goods with a chemical odor are returned to you on subsequent visits, try a different dry cleaner.

## Indoor Air Pollutants

There are many indoor air contaminants, which can be separated based on their effects on human health, the frequency of their appearance, their usual concentration levels, their sources etc.

### Radon

The main source of indoor radon is its immediate parent radium-226 in the ground of the site and in the building materials. Outdoor air also contributes to the radon concentration indoors, via the ventilation air. Tap-water and the domestic gas supply are usually radon sources of minor importance, with a few exceptions. In most situations it appears that elevated indoor radon levels originate from radon in the underlying rocks and soils. This radon may enter living spaces in dwellings by diffusion or pressure driven flow if suitable pathways between the soil and living spaces are present. It should be noted, however, that in a minority of cases elevated indoor radon levels may arise due to the use of building materials containing high levels of radium-226. Examples of such materials, used in some buildings, are by-product gypsum, alum shale and volcanic tuffs.

The United Nation Scientific Committee on the Effects of Atomic Radiations (UNSCEAR) has made a very simple model to try to estimate the relative contribution of these sources: for a "typical" house, with a radon concentration of 50 Bq/m³ at ground floor, the contributions of soil, building materials and outdoor air are, respectively, 60%, 20% and 20%, while for the upper floors in high rise buildings, where the radon concentration-is estimated to be "typically" 20 Bq/m³, these values become 0%, 50% and 50%.

### Soil

For those who live close to the ground, e.g. in detached houses or on the ground floor of apartment buildings without cellars, the most important radon source is radium in the ground.

The radium concentration in soil usually lies in the range 10 Bq/kg to 50 Bq/kg, but it can reach values of hundreds Bq/kg, with an estimated average of 40 Bq/kg. Typical radon concentrations in soil gas range from 10000 Bq/m³ into 50000 Bq/m³. The potential for radon entry from the ground depends mainly on the activity level of radium-226 in the subsoil and its permeability with regard to air flow. Example of terrains with a high radon potential are alum shales, some granites and volcanic rocks, due to high concentrations of radium-226 and the presence of eskers (gravel, sand and rounded stone deposited from subglacial streams during the ice ages), all these being characterised by high permeability. The ground could also be contaminated with waste tailings from uranium or phosphate mining operations with enhanced activity levels.

The ingress of radon from the soil is predominantly one of pressure-driven flow, with diffusion playing a minor role. The magnitude of the inflow varies with several parameters, the most important being the air pressure difference between soil air and indoor air, the tightness of the surfaces in contact with the soil on the site, and the radon exhalation rate of the underlying soil. If there is no airtight layer between the basement and the ground, the underpressure indoors causes radon to be drawn in from the ground under the building. Underpressure occurs in most houses if either the adjustment of inlet and outlet of air in forced ventilation systems or the outdoor air supply for vented combustion appliances is inappropriate. The underpressure may be considerable for all types of ventilation systems when the inlet air is restricted too much. The tightness of the structures has to do with e building regulations and techniques and is very dependent on cracks, openings and joints. Structures are hardly ever so airtight that radon inflow is completely prevented. For example, to get a radon daughter concentration of less than 100 $Bq/m^3$ EER in a house with a volume of 500 $m^3$ and a ventilation rate of 0.5 air changes per hour, not more than 1 $m^3$ per hour must be allowed to leak into the house if the radon gas concentration in soil air is about 50000 $Bq/m^3$. Such values are quite typical.

## Building Materials

Building materials are generally the second main source of radon indoors, while in the Seventies they were considered the principal one. Radon exhalation from building materials depends not only on the radium concentration, but also on factors such as the fraction of radon produced through material release, the porosity of the material and the surface preparation and finish of the walls. In general, no action needs to be taken concerning traditional building materials. Typical values for radium and thorium content in building materials are 50 Bq/kg or less. Building materials containing by-product gypsum and concrete containing alum shale may have much higher radium concentrations. The activity concentrations in brick and concrete may also be high if the raw materials have been taken from locations with high levels of natural radioactivity. Examples of such natural materials, used in some buildings, are volcanic tuffs and pozzolana, where radium and thorium content can reach some hundreds of Bqlkg. Other measurements of radioactivity content and exhalation of building materials are reported in NENOECD.

Building materials are the main sources of radon-220 (also called "thoron") in indoor air. Due to its short half life (55 s), thoron originating in soil in effect is usually prevented from entering buildings and therefore makes negligible contribution to indoor thoron levels. For this reason and due to the greater difficulties of measurement, thoron concentration measurements are very much fewer than those for radon. Although the indoor thoron concentrations are usually low, in some cases the doses due to this isotope and its daughters are significant and comparable to those due to radon- 222.

## Outdoor Air

Outdoor air usually acts as a diluting factor, due to its normally low radon concentration, but in some cases, as in high rise apartments built with materials having very low radium content, it can act as a real source. The radon concentration in outdoor air is mainly related to atmospheric pressure, and (in case of non-perturbative weather) it shows a typical oscillating time pattern, with higher values during the night.

Until a few years ago the average level of radon gas concentrations in the atmosphere at ground level was, in most cases, assumed to be of the order of few $Bq/m^3$ -e.g. in the range of 4 to 15 $Bq/m^3$ in USA, but more recent measurements seem to indicate higher values, reaching some tens of $Bq/m^3$. Quite high radon concentrations in the outdoor air have been reported near substantial radon sources, such as mine tailings, or in the case of particular weather conditions, such as thermal inversion or very low precipitation.

Ambient air over oceans has very low values ($\sim 0.1$ $Bq/m^3$) of radon concentrations, due to the minimum presence of radium in the sea water and the high solubility of radon in water at low temperatures. Therefore radon concentration in outdoor air of islands and coastal regions is generally lower than in continental countries, e.g. United Kingdom and Japan have an average outdoor air value of $\sim 4$ $Bq/m^3$.

Taking into account recent measurements, the mean value of outdoor radon concentrations adopted by UNSCEAR in its last report has been changed from 5 to 10 $Bq/m^3$ for continental areas and somewhat less in coastal regions.

## Tap Water

In wells drilled in rock the radon concentrations of water may be high. When such water is used in the household, radon can be partially released into the indoor air, causing an increase in the average radon concentrations. In a few regions, such as Finland and Maine (USA), the tap water from wells drilled in rock has been shown to contribute significantly to radon concentrations indoors. Radon concentrations in tap-water from deep wells can range from 100 $kBq/m^3$ to 100 $MBq/m^3$. The indoor radon concentrations in these regions may already be high due to high rates of radon entry from the ground. The world average radon concentration in all types of water supplies is assumed to be 10 $kBq/m^3$.

## Domestic Gas

In some regions, natural gas used for cooking and heating contains elevated concentrations of radon, which is released on combustion. Normally this source is insignificant, and can be monitored at transmission and distribution points. Typically the radon level in natural gas is about 1000 $Bq/m^3$. Natural gas, as it is usually supplied, contains gas from a number of wells and fields and thus can vary over time, depending on the proportions supplied by different sources.

## Oxides of Nitrogen

### $NO_x$

A large number of studies of NO and $NO_2$ have been carried out in many different indoor air environments. Because of air exchange, indoor levels are generally higher when outdoor levels increase. However, enhanced indoor levels can be found when combustion sources are present. These include gas stoves, paraffin heaters, water heaters, and cigarette smoke. While combustion generates primarily NO, the focus indoors has been on $NO_2$ because of its health impact. Again, the use of gas stoves was highly correlated with indoor $NO_2$, with an indoor/outdoor concentration ratio of 1.19 for homes with a gas range compared to 0.69 for those without a gas stove. The ratio was even higher for homes with a paraffin space heater, 2.3 compared to 0.85 without such a heater. Both the indoor and outdoor concentrations of $NO_2$ were higher in cities where at least 75% of the homes had gas stoves; for example, the mean outdoor $NO_2$ concentration in such gas-intensive cities was 38 ± 20 ppb, compared to 14 ± 6 ppb in cities where fewer than 25% of the households had gas stoves installed. High concentrations of $NO_2$ have also been measured in indoor skating rinks where the use of ice resurfacing machines powered by propane, gasoline, or diesel fuel results in significant emissions. Mean concentrations of $NO_2$ of ~200 ppb have been reported, with some rinks having concentrations up to 3 ppm! The indoor-to-outdoor ratios of the arithmetic mean concentrations varied from about 1 to 41, with an overall mean of 20. In the absence of such sources of $NO_x$, indoor and outdoor concentrations are quite similar, since removal of NO and $NO_2$ indoors, e.g., on surfaces, is relatively slow. However, as it has been discussed shortly, although the surface reaction of $NO_2$ is relatively slow, it is still of interest since it generates nitrous acid (HONO). Different surfaces found inside homes have been found to have different removal rates for $NO_2$. In short, there is a variety of evidence that there are higher levels of $NO_2$ indoors when combustion sources are present and that the concentrations generated indoors can be quite substantial in some circumstances. One word of caution is in order, however, particularly in regards to earlier measurements of $NO_2$.

## HONO and $HNO_3$

HONO is formed by the reaction of $NO_2$ with water on surfaces. The reaction is usually represented as:

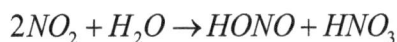

$$2NO_2 + H_2O \rightarrow HONO + HNO_3$$

Although the detailed mechanism is not known; gaseous $HNO_3$ is not generated in equivalent amounts, something which has been attributed to its remaining being adsorbed on the surface. This overall reaction occurs on a variety of surfaces in the laboratory and hence might be expected to also occur on surfaces in other environments, such as homes. This, indeed, is the case. first used differential optical absorption spectrometry (DOAS) to establish unequivocally that $NO_2$ injected into a mobile home forms HONO. Interestingly, the dependence of the rate of HONO generation on the $NO_2$ concentration was similar to that measured in laboratory systems, consistent with production in, or on, a thin film of water adsorbed on surfaces. A number of studies have confirmed that the behaviour is similar to that in laboratory systems; i.e., the rate of production of HONO increases with $NO_2$ and with relative humidity. Indoor levels of HONO as high as 8 ppb as a 24-h average and 40 ppb as a 6-h average have been reported in normal, in-use buildings and homes. The ratio of HONO to $NO_2$ indoors can be quite large, up to ~0.15. This can be compared to typical values of a few percent outdoors. High levels of HONO (up to ~ 30 ppb) have also been measured in automobiles in use in polluted urban areas, and again, the ratio of HONO to $NO_2$ was quite large,

~0.4, compared to 0.02–0.03 measured outdoors in the same study. The generation of NO was attributed by Spicer and co-workers to a reaction of gaseous $NO_2$ with adsorbed HONO:

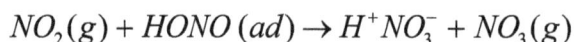

$$NO_2(g) + HONO\,(ad) \rightarrow H^+NO_3^- + NO_3(g)$$

The same process was hypothesised to explain some time periods in a commercial office building when indoor NO actually exceeded outdoor NO. As is the case in laboratory systems, equivalent amounts of $HNO_3$ are not observed as might be expected from the stoichiometry of reaction, likely due to $HNO_3$ remaining on the surface after formation and/or being taken up by surfaces. The accumulation of nitrate on indoor surfaces in a commercial building has been reported by Weschler and Shields and attributed to the formation and uptake of $HNO_3$ via reactions of $NO_3$ and/or oxidation of nitrite (i.e., adsorbed HONO) in an aqueous surface film. Subsequently, it was shown that HONO is also directly emitted by gas stoves. In a house used for investigating indoor air pollution that had natural gas fueled appliances (a convective heater, a radiant heater, and a range with four burners), both the surface reaction of $NO_2$ and the direct combustion emissions contributed significantly to the measured indoor HONO. When an appliance was operational, the contribution of direct emissions was the more important source. In short, the "dark reaction" of $NO_2$ with water on surfaces is ubiquitous and occurs not only in laboratory systems but also indoors. The combination of this heterogeneous reaction with combustion sources of HONO can produce significant concentrations of HONO indoors. As a result, there is a concern regarding the health impacts of nitrous acid, not only because it is an inhalable nitrite but also because it is likely the airborne acid present in the highest concentrations indoors.

## Microbial Pollutants

Microbial pollution is a risk to health and is associated with allergic illnesses. Published results indicate that 20% of the population can be sensitised by airborne fungal spores in the UK, while 40% of the inspected houses in Germany suffer from mould-related problems. The medical consequences of immune response, allergic reactions, endotoxins, mycotoxins, and epidemiology have been extensively studied by Miller, Morey, Gravensen et al. and Burge et al. Similarly, Legionnaires' disease and Pontiac fever are associated with wet cooling towers and domestic hot-water systems in complex buildings.

Accordingly to the official published figures, some 560,000 people need treatment because of indoor pollution due to mites and mould in damp houses. Indoor airborne allergic components come from two sources: outdoor air-borne spores moving inside and allergic components originating inside the dwelling. The source of biological growth within buildings is associated with moisture and the formation of microclimates; it also depends upon the type of the buildings and their ventilation. Mould fungi thrive on surfaces on which there is nourishment and suitable humidity, for example on damp water pipes, windows and walls in kitchens and bathrooms, in central air-conditioning systems, circulation pumps, blowers, ventilation ductwork and air filters, central dehumidifiers, and inside damp structures. Allergenic substances can be airborne and inhaled, such as pollen, fungus and dust, digested, such as mouldy food or drink. Investigations suggest that airborne allergies cause more problems throughout the world than all other allergies combined. Additionally, cross-infection from patient to patient is of great concern in hospitals. The medical field that treats allergies recognises the following allergenic diseases: asthma, allergic rhinitis, serous otitis media, bronchopulmonary aspergillosis, and hypersensitivity pneumonitis.

Allergic load and cocktail effect: For some people, an allergic reaction in the indoor environment may be triggered by non-biological factors, such as chemicals or other indoor air pollutants, emotional stress, fatigue or changes in the weather. These factors burden allergic people further if they are suffering from allergic reactions to biological contaminants. This combination is known as 'allergic load'. Microbial contaminants propagated within the health care establishment are particularly aggressive to patients due to reduced immune system resistance.

Recently, attention has been focused on the cocktail effect of chemicals present in indoor air. Volatile organic compounds may be produced from the use of wood preservatives and remedial timber treatment chemicals, moth-proof carpets, fungicides, mouldicide-treated paints, furnishing materials such as particle board and foamed insulation which may emit formaldehyde. Biological pollutants alone or in synergetic effect with any of the above-mentioned volatile organic compounds may produce symptoms such as stuffy nose, dry throat, chest tightness, lethargy, loss of concentration, blocked, runny or itchy nose, dry skin, watering or itchy eyes or headache in sensitive people. The 'sick building syndrome' (SBS) or tight building syndromes may arise from a variety of causes. Because of the uncertainties about the causes of SBS and the rising levels of health related problems in buildings there is an increasing use of the term building-related illness (BRI) to cover a range of ailments which commonly affect building occupants.

## Asbestos and Manmade Mineral Fibres

Asbestos is known to cause a number of diseases after occupational exposure. Before the hazards associated with the inhalation of these mineral fibres were understood these exposures were often very large with frequent reports of dust clouds so great that visibility in the workplaces was considerably reduced. This type of exposure is quantitatively quite different from those in the general environment that have provoked a response which in some quarters approaches hysteria. In the USA at least there is massive expenditure on asbestos removal, management and litigation.

Asbestos is a collective, trivial, name given to a group of highly fibrous minerals that are readily separated into long, thin, strong fibres occurring on sufficient large bulk deposits for their industrial exploitation. Asbestos minerals were usually used for their insulating properties, or in a composite, where they added strength, as in cement, or increased friction, as in brake shoes. Chrysotile, or white asbestos has counted for over 90% of the world trade in asbestos minerals. It is a serpentine mineral while the others (amosite (brown asbestos); crocidolite (blue asbestos); anthophyllite; tremolite; and actinolite) are all amphibole minerals. Amphibole asbestos has grater acid and water resistance than chryusotile and was used where these properties made it more suitable. Sometimes users would be unaware of the differences between the types of asbestos and so different minerals could have been used for a single application.

Recently the concern over the health effects of asbestos has been extended to another group of fibrous materials- the man-made mineral fibres (MMMF). While this term is self-explanatory a variety of types are produced with diverse chemical compositions, properties and uses. While sometimes referred to as 'asbestos substitutes' the majority of uses for the manmade fibres are relatively novel and ones for which the natural fibres are unsuitable. For example refractory ceramic fibres are resistant to considerably higher temperatures than are any of the natural fibres. The development of synthetic fibrous insulation materials has been given a great impetus in recent years by the need for more thermally efficient buildings and industrial processes.

MMMF can be made from most types of glass, from rock such as basalt, diabase and olivine and from various types of slag. Ceramic fibres can be made from kaolin or from pure silica and other oxide starting materials. The MMMF have been classified into four broad groups based on the manufacture and use: continuous filament glass fibre made by extrusion and winding processes, insulation wool (including ceramic fibre), and special purpose fibres. The non-continuous fibres are made by dropping molten material onto spinning disks or by air or steam jet impingement on a stream of the molten material. They contain a wide range of fibre sizes and are contaminated by small glassy balls called shot which often account for 50% of the product by weight.

## Factors that Influence Exposure to Indoor Air Pollutants

### General

The types and quantities of pollutants found indoors vary temporally and spatially. Depending on the type of pollutant and its sources, sinks and mixing conditions, its concentration can vary by a factor of 10 or more, even within a small area.

Human mobility constitutes an important kind of complexity in the determination of exposure to air pollutants. Human activity patterns differ between midweek and weekend, between one season and another, and between one part of one's life and another. Activity patterns determine when and how long one is exposed to both indoor and outdoor pollutants. Therefore, in reviewing the factors that influence air-pollution exposures, we have specifically separated them into two major components: time (activity) and concentration (location).

Information on the time spent in various activities is summarised first, and then the variations in concentration often encountered in different locations. Unfortunately, most of the studies discussed were not longitudinal and thus do not offer information on seasonal differences in time spent indoors and outdoors or on regional differences in activity patterns.

Outdoor concentrations of pollutants and rates of infiltration affect the concentrations to which people are exposed indoors. Building construction techniques, as they vary geographically, and their effect on pollution infiltration are particularly important. But the measurement techniques available are limited; the need for additional studies is discussed. The rates of infiltration on a neighbourhood scale have been studied by only a few researchers. Although their work has focused on energy conservation, their findings can easily be applied to the study of impact on indoor pollution.

Patterns of human behaviour and activity determine the time spent in any specific location, and thus knowledge of them is essential in estimating exposures of populations to pollutants. As indicated by Ott, a large number of variety of studies in which data on human activities were collected from population samples have been completed over the past 50 ye.

When one examines the literature on human activities, the term "time budget" ("zeitbudget", "budget de temps") is encountered often. A time budget produces a systematic record of how time is spent by a person in some specified period, usually 24 h. It contains considerable detail on a person's activities; including the locations in which the activities take place.

One way of obtaining time budget information from the populations surveyed is to ask each respondent to maintain a diary of his or her activities over a 24-h period or longer. In another approach, the so-called "yesterday" survey approach, the interviewer asks each responder about his or her activities on the preceding day.

Several summaries of the historical development of time-budget research have been published discussed the literature on activity patterns in the context of estimation of exposure to air pollution. Owing to the small number of field monitoring studies, the geographic distribution of indoor air pollutants has not been determined. However, it is instructive to review the geographic distribution of the major factors that affect variations in the concentrations of pollutants and their impact on the quality of the indoor environment. Outdoor air quality, air-infiltration rates, and sources of emission of indoor air pollutants are the major factors. Outdoor air quality has been studied with respect to some pollutants, and the geographic distribution of these few pollutants is well understood. Descriptive statistics published annually by EPA and state and local air-quality agencies furnish much scientific information useful in discerning regional and local differences in concentrations of carbon monoxide, total suspended particles, ozone, $NO_x$, sulfur dioxide, sulfates, and others. It should be noted that the geographic distribution of some criteria pollutants has been studied and is easily accessible from the literature; information on non-criteria pollutants is sparse and often collected and analyzed by questionable methods.

Concentrations of chemically non-reactive pollutants in residences generally correlate with those outdoors. Distribution of indoor air quality is extremely difficult to describe on a geographic scale, because indoor air quality is determined by complex dynamic relationships that depend heavily on occupant activity and highly variable structural characteristics. Weather, which has a regional character, influences indoor air concentrations of some chemicals, such as formaldehyde, and biologic contaminants, such as bacteria and molds. Therefore, the influence of relative humidity and other weather-related conditions affecting indoor environmental quality needs to be studied geographically. Research specifically addressed to geographic distribution of indoor air quality is needed.

Typically, the air-infiltration rate for American residences is assumed to be 0.5- 1.5 ach. This assumption is supported by the results of several energy and air-quality studies that experimentally determined the range of ventilation rates for typical residences to be between 0.7 and 1.1 ach. However, the sample that yielded the data is small, and statistical documentation for such statements is not strong.

The quality of indoor air is a function of outdoor air quality, emission from indoor sources, air-infiltration rates, and occupant activity is likely to vary within each metropolitan and suburban area, is indeed within each neighbourhood. Within a metropolitan area, it has been shown that an urban complex leads to the so-called urban heat reservoir (American Society of Heating, Refrigerating and Air-Conditioning engineers. ASHRAE). Urban characteristics- such as city size, density of buildings, and population- correlate with such meteorological factors as temperature, pressure and wind velocity. The urban heat island affects both urban pollution patterns and meteorological characteristics that affect the infiltration rates of buildings. Thus, although the exact nature of the impact on indoor air quality is not known, it is fair to expect that the heat island to have an impact on the indoor environment that is likely to be adverse. Also, the variations due to mechanical ventilation, structural differences, and air infiltration may vary within a neighbourhood as a function of such factors as house orientation, tree barriers, and terrain roughness.

Occupant activity, air-infiltration rates, the indoor sources of pollutants and their chemical natures are some of the factors that cause variations within a city. A study in the Boston metropolitan area obtained indoor air samples from 14 residences under occupied "real-life" conditions for 2 week each. The indoor air character not only was driven by outdoor concentrations, but was greatly affected by other factors, such as indoor activities.

Wind speed, temperature difference, pressure differential, terrain characteristics (roughness and barriers, such as trees and fences), building orientation, and structure characteristics may be affected by the location of one residence relative to another within a neighbourhood.

The indoor air quality of an individual building is often characterised by the 24-h average for the concentration of one pollutant measured at one sampling location. Because the activity patterns of persons are such that more time is spent in some indoor areas than in others, the question arises "Do indoor zones (independent areas) with distinct pollutant patterns exist?" At issue here is whether sampling from one monitoring zone is sufficient to characterise the air quality of an entire building.

In an extensive analytic study of indoor air quality, Shair and Heitner assumed that there are no pollutant gradients in the indoor environment. The experimental database of Moschandreas and co-workers verified that the gradients in concentrations of several gaseous pollutants in the residential environment are negligible. J. D. Spengler, R. E. Letz, J. B. Ferris, Jr., T. Tibbets, and C. Duffy reported on weekly nitrogen dioxide measurements in 135 homes in Portage, Wisconsin. On the average, kitchen concentrations were twice those in bedrooms in homes that had gas stoves. A study of the air quality in a scientific laboratory by West showed an almost uniform distribution of an intern tracer continuously released in the room. Similar experiments performed by Moshandreas et al. in residential environments showed that equilibrium is reached throughout a house within an hour. Episodic release of sulphur hexafluoride tracer gas also illustrates this point. The source location was the living room; adjacent locations were the kitchen and the hall. Episodic release of this inert gas in 24 residences was followed by uniform indoor distributions within 30 min. The one-zone concept does not require instantaneous mixing, because it is based on the behaviour of hourly average pollutant concentrations.

Moschandreas and associates used a different database derived from the monitoring of 14 indoor environments in the Boston metropolitan area. Analysis of variance was used to reach the following conclusions:

- Pollutants (ozone and sulphur dioxide) generated principally outdoors have little or no interzonal statistical difference indoors.

- Pollutants with strong indoor generation have interzonal statistical differences in residences with gas facilities and offices, but not in electric-cooking residences. In general, the observed differences are not large, and the health differences are not expected to be serious.

- Depending on indoor activity and outdoor episodic pollutant activity, the indoor arithmetic 24-h average may or may not adequately represent the variation of hourly indoor concentrations.

- Although more than one zone would be preferable, hourly pollutant concentrations obtained from one indoor zone adequately characterise the indoor environment.

The most important factors that influence exposure to indoor air pollutants are the one described under. It should be noticed that these conclusions are not applicable to short-lived pollutants. Contaminants associated with tobacco smoke, bathroom odours, allergens, and other pollutants related to dust are expected to vary considerably in a given residence. Additional documentation is needed to determine the extent of this variation.

## Site Characteristics

The characteristics of a building site that influence indoor air quality are addressed as three related subjects: air flow around buildings, proximity to major sources of outdoor pollution, and type of utility service available.

The air flow around a building has been shown to be determined by the local characteristics of the geometry of surrounding buildings, the location and type of surrounding vegetation, the terrain, and the size and shape of the building itself. Pollutants can be transferred by the air flow from the street level, over the façade of the building and onto the roof. Field tests of isolated buildings have been used to develop scaling coefficients for both isothermal and and stratified cases of surface wind pressures, turbulence, and dispersion. Air flow around the building creates low pressure on the leeward side and/or the sides adjacent to the windward face, as well as the roof. Air pollutants released from stacks, flues, vents, and cooling towers in the region can re-enter the building through make-up air intakes for ventilation.

Trees and forests have been generally studied as shelter belts in an agricultural context. Shelter belts affect air flow around buildings. When an air current reaches a shelter belt, part of it is deflected upward with only a slight change in velocity, part passes through the crowns of the trees with very low velocity, and part is deflected beneath the canopy with rapidly decreasing velocity. The changes in velocity of air flow outside may change the infiltration rate and thus affect indoor air quality.

The location of a building relative to a major outdoor pollution source can affect indoor air quality. For example, buildings near major streets or highways often have high carbon monoxide and lead concentrations, owing to the infiltration of these pollutants.

The type of utility service available is also related to the site of the building and may affect the character of its indoor environment. The availability of particular fuels (e.g., natural gas and oil) influences the types and concentrations of pollutants (e.g., combustion products) emitted by space- and water- heating. Service moratoria, development timing, and development scale are institutional elements that contribute to the variability of utility services and thus can affect indoor air quality.

## Occupancy

Occupancy factor that affect indoor air quality include the type and intensity of human activity, spatial characteristics of a given activity, and the operation schedule of a building.

Several human activities-such as smoking, cleaning and cooking- generate gaseous and particulate contaminants indoors. The number of occupants of a space and the degree of their physical activity (i.e., metabolic rate at rest or under intense activity) are related to the production of various pollutants, such as carbon dioxide, water vapour, and biologic agents. If the only source of indoor carbon dioxide is that caused by occupants, ventilation rates may be proportional to the number of people and their metabolic rates. Although studies have shown no constant relationship between carbon dioxide concentrations and the concentrations of other pollutants, carbon dioxide concentration is often used as a general indicator of the adequacy of ventilation in an occupied space.

Building occupancy is often expressed as occupant density and the ratio of building volume to floor area. The importance of occupancy in indoor air quality is illustrated by the fact that the choice of natural or mechanical ventilation is based on occupant density and the spatial characteristics of the building under consideration.

Occupancy schedule and associated building use may affect the type, concentration, and time and space distribution of indoor pollutants. Because most buildings are unoccupied for substantial portions of each day, the manipulation of "operating schedule" is a means of controlling energy use. Efforts to conserve energy through the design of ventilation systems can result to the degradation of indoor air quality. However, detailed studies relating ventilation capacity, occupancy schedules, energy requirements, and indoor air quality have only recently been implemented.

## Design

Elements of building design that affect the indoor environment include interior-space design (space planning), envelope design, and selection of materials.

The evolution of space planning in many building types has resulted in flexibility in assigning functions to specific locations. However, this flexibility is accompanied by a decrease in the ability to predict exposure to air pollutants. In particular, "open-plan" offices and schools have serious technical problems of redundant service distribution, limited acoustic control, incomplete air diffusion, and incomplete pollutant dispersion indoors, compared with "fixed-plan" floor layouts.

Evaluation of the success of a floor plan in achieving space efficiency, structural economy, and energy efficiency is usually in terms of net area per occupant and ratio of net usable area to total area. Explicit planning for environmental quality must be included to ensure that spatial arrangements are acceptable to the occupants.

A building's structural envelope consists of both primary elements -foundations, floors, walls, and roofs- and secondary "skin" elements -facings, claddings, and sheathing. To various degrees, the function of these is to maintain the integrity of the structure under the stresses caused by structural load, wind pressure, thermal expansion, precipitation, earth movement, and fire. The integrity of the building envelope is a major consideration in uncontrolled air movement into and out of the building —usually referred to as "infiltration". This is a major factor in indoor air quality. There has been no systematic survey of infiltration rates of buildings in the United States. The dominant factor in determining a building's infiltration rate is the total area of effective leakage, as measured with fan pressurisation. Following the leakage area in importance are the terrain and shielding near the building, the mean climatic conditions during heating (or cooling) periods, and the building height. There is much evidence, both in the United States and in Europe, that houses in mild climates are "very leaky", whereas houses in severe climates are "tight".

Greater height of a building increases the "stack effect", or updraught, and exposes the building to higher wind speeds. Thus, higher wind pressures drive air through existing openings, referred to as "leakage", increasing the infiltration rate.

The dominant building factors that determine infiltration have not been identified, but a catalogue of leakage openings found in typical structures is as follows:

- Walls: Leakage around sill plates (the openings at the bottom of wallboard), electric outlets, plumbing penetrations, and headers in attics for both interior and exterior walls.

- Windows and doors: Window type is more important than manufacturer in determining window leakage. This source of leakage tends to be overrated; it contributes only about 20% of the total leakage of a house.

- Fireplaces: This includes dampers, glass screens, and fireplace caps.

- Heating and cooling systems: The variables include combustion air for furnaces, dampers for stack air draft, air-conditioning units, and location of ductwork.

- Vapour barrier and insulation penetrations.

- Utility accesses: This includes recessed lighting and plumbing and electric penetrations leading to attic or outside.

- Terminal devices in conditioned space: This includes leakage of dampers, especially those for large air-handling systems.

- Structural types: Examples are drop ceilings above cupboards or bathtubs, prism-shaped enclosures over staircases in two-story houses, and elevator and utility shafts that lead from basement to attic.

Wall and ceiling materials and floor finishes are the constituents of the building interior. Modular components, weight, strength, thermal insulation, thermal stability, sound insulation, fire resistance, ease and speed of installation and ease of maintenance are among the criteria considered in the selection of materials for walls, ceiling and floors. But emphasis on first cost, ease of installation, maintenance and long service life has also led to the use of materials that may be sources of indoor contaminants.

## Operations

Depending on the type of ownership (owner-occupied or developer-owned), building operation may vary considerably, and this variation may have an impact on indoor air quality. "Building operation" pertains to the following elements of a building: the building envelope, service and plant, building facilities, equipment and landscaping. Cleaning, preventive maintenance, and replacement and repair of defects are also included in building operation. The staff responsible for building operation includes management, engineering, and custodial personnel. The care responsibilities are operation of the heating, ventilation, and air-conditioning systems and building services, such as hot water, lighting and power distribution. Building operation has an impact on indoor air quality in numerous ways, but the magnitude of this impact is not known.

## Health Effects of Indoor Air Pollution

Indoor air pollution, apart from the health impact, has socio-economic costs. The potential economic impact of poor indoor air quality is quite high, and has been estimated to be in the order of tens of billions of ECU per year in Western Europe. This includes costs of medical care, loss of income during illness, days lost due to illness, poor working performance and lower productivity. Labour costs are significantly greater per square metre of office space than energy and other environmental control costs. In the US, the loss in productivity for each employee which is attributable to IAQ problems is currently estimated to be 3% (14 minutes/day) and 0.6 added sick days annually. Other estimates have been made by calculating the impact of IAQ on productivity. For instance, in Norway, the authorities estimate that the costs to society related to poor IAQ are in the order of 1 to 1.5 billion ECU per year or about 250 - 350 ECU per inhabitant. This estimation only includes costs related to adverse health effects requiring medical attention and does not include reduced working efficiency or job-related productivity losses. Thus, from an economic consideration, remedial action to improve indoor air quality is likely to be cost effective even if an expensive retrofit is required.

As far as it concerns the health effects on IAP, it is very interesting to present the methods of studying health effects, the criteria for the assessment of the impact of IAP on the community and the diverse effects of IAP on human health.

## Methods of Studying Health Effects

Methods of studying health effects of indoor pollutants can be grouped into three broad categories. Human studies, subdivided into observational and experimental studies:

a. Epidemiological studies of pollutants are mostly observational, i.e. the investigator has no means of experimentally exposing humans to pollutants, or of allocating subjects to exposed and unexposed groups. Critical issues are therefore the validity and precision of exposure assessment, and the control for confounding factors in these studies. Recent developments have stressed the importance of reducing exposure misclassification, and of studying restricted, well defined, homogenous populations to address these issues. The main advantage is that humans are studied under realistic conditions of exposure. By themselves, observational epidemiological studies are not usually sufficient to support causality of an observed association, so that additional information is needed from other types of studies. Experimental studies are among these; however, these are only suitable for studying moderate, reversible, short term effects in persons who are healthy or only moderately ill. Their main advantage is that exposure conditions and subjects election are under the control of the investigator.

b. Animal studies, which can be subdivided into a number of categories depending on their length (acute, subchronic, chronic) or end-point (morbidity, mortality, carcinogenicity, irritation, etc.). Here, the investigator has full control over exposure conditions and health effects studied. However, the principle limitations lie in the fact that extrapolation from the studied animal species to man is always necessary. Also, while in human populations health effects with low incidences are often of interest (e.g., specific cancers), it is not feasible to study very large groups of animals to detect these low incidences. In practice, therefore, animal experiments are often carried out using very high experimental doses to compensate for the relatively small number of animals used and as a consequence, an additional extrapolation from high to low doses is also often necessary.

c. In vitro studies, in which effects of pollutants on cell or organ cultures are studied. These studies have the advantage that they are less costly than animal studies, and that results can generally be obtained in a shorter period of time. They are useful for studying mechanisms of action, but it is not usually possible to predict effects on whole organisms from their results in a quantitative way.

## Criteria for the Assessment of the Impact of IAP on the Community

The process of risk characterisation for indoor pollutants occurs through several phases: hazard identification, exposure assessment, dose-effect evaluation, and finally qualitative and quantitative risk assessment. The final product of this process may be an individual risk estimate per exposure unit or the evaluation of the incidence of the concerned effects in a given population. The risk characterisation through a multi-stage process as described above is particularly informative because, by dividing the analysis of the scenario of each pollutant into steps, it allows the separate recognition of the importance of each variable in the scenario and the prediction of the changes of frequency or severity of effects obtainable by modifying (increasing or decreasing) exposure.

For some types of IAP, our understanding of human health risk is well defined. For most indoor air pollutants, however, the risk assessment process has its limitations.

First, it has been applied successfully only to individual pollutants for which information is available for exposure and dose-response relationships and for which the effect is clear, certain, and measurable, such as mortality and cancer. Little progress has been made in applying the risk assessment process to environmental issues involving pollutant mixtures or effects for which the causes are difficult to ascertain precisely, such as in heart disease, allergic reactions, headache, and malaise. A different approach is needed for the assessment and characterisation of the risks associated with most indoor air pollutants.

A basic and simple criterion for assessing the importance of the health risk related to indoor pollution makes reference to the severity of the effect concerned and to the size of the population affected. Important issues for the community may come from severe health impacts, particularly when affecting a large segment of the population. Minor impacts, such as those related to discomfort or annoyance may, however, become important when a large number of individuals in the community are concerned.

## The Impact of IAP on Humans' Health

## Respiratory Health Effects Associated with Exposure to IAP

Several effects on the respiratory system have been associated with exposure to IAP. These include acute and chronic changes in pulmonary function, increased incidence and prevalence of respiratory symptoms, augmentation of pre-existing respiratory symptoms, and sensitisation of the airways to allergens present in the indoor environment. Also, respiratory infections may spread in indoor environments when specific sources of infectious agents are present, or simply because the smaller indoor mixing volumes allow infectious diseases to spread more easily from one person to the next. The latter mechanism is particularly operative in schools, nursery schools, etc.

Observed changes in pulmonary function due to exposure to, e.g., tobacco smoke in the home, have mostly been due to acute or chronic airway narrowing leading to obstruction of air flow. This is measured as a reduction in the quantity of air that can be exhaled in one second after deep inspiration (FEVI), and a limitation in the various measures of air flow such as Peak Expiratory Flow (PEF), Maximum Mid Expiratory Flow (MMEF), and Maximum Expiratory Flow at x% of Forced Vital Capacity (MEFx). In growing children, it has also been suggested that lung development could be impaired by exposure to IAP.

Asthma, manifested by attacks of excessive airway narrowing leading to shortness of breath and wheezing, can be caused or aggravated by exposure to allergens at home, but it has also been associated with exposure to substances such as nitrogen dioxide and environmental tobacco smoke (ETS). Bronchitis, manifested in inflammatory changes in the airways and mucus hypersecretion has been linked to high levels of ambient air pollution in the past, and to exposure to ETS in the home in recent studies. Respiratory symptoms which have been associated with exposure to indoor air pollutants are symptoms mostly related to the lower airways such as cough, wheeze, shortness of breath and phlegm.

In contrast to the occurrence of chemical pollutants in indoor air, attention to which has grown considerably over the past two decades, the role of infectious agents in indoor air has been known for a long time. Infectious agents can be involved in the inflammatory conditions rhinitis, sinusitis, conjunctivitis and sinusitis, in pneumonia, in asthma and in alveolitis.

## Allergic Diseases Associated with Exposure to IAP

Allergic asthma and extrinsic allergic alveolitis (hypersensitivity pneumonitis) are the two most serious allergic diseases caused by allergens in indoor air. Allergic rhinoconjunctivitis and humidifier fever are other important diseases; it is not clear if or how the immunological system is involved in humidifier fever.

Allergic asthma is characterised by reversible narrowing of the lower airways. Pulmonary function during an attack shows an obstructive pattern in serious cases together with reduced ventilation capacity. Allergic asthma may be caused by exposure to indoor air pollutants, either acting as allergens or as irritants. Immunological specific IgE sensitisation to an airborne allergen is a major component of this disease, but non-specific hypersensitivity is also important for the asthmatic attacks occurring on exposure to irritants in the indoor air.

The prevalence of asthma varies considerably from country to country. Although asthmatic attacks seldom lead to death, the costs of medical care are considerable in terms of hospital admissions, medication, and lost work days.

Allergic rhinoconjunctivitis is also an IgE-mediated disease, but while asthma occurs in all age groups, allergic rhinoconjunctivitis is especially prevalent among children and young adults. The main symptoms are itching of the eye andlor the nose, sneezing, watery nasal secretion and some stuffiness of the nose. The severity of the symptoms varies with the exposure to the allergen. Individuals often suffer from both allergic asthma and allergic rhinoconjunctivitis and are seldom sensitive to only one allergen. Aeroallergens from house dust mites, pets, insects, moulds, and fungi in the indoor air have been shown to be associated with allergic asthma and rhinoconjunctivitis. Extrinsic allergic alveolitis, also called hypersensitivity pneumonitis, is characterised by recurrent bouts of pneumonitis or milder attacks of breathlessness and flu-like symptoms. Studies of the pulmonary function during an acute episode will usually show a restrictive pattern with a decreased diffusion capacity. The disease is believed to be an inflammatory reaction in the alveoli and bronchioles involving circulating antibodies and a cell-mediated immunological response to an allergen. For example it occurs in farmers as a result of handling mouldy hay ("farmer's lung") and in pigeon breeders due to bird droppings. However, the disease has also in a few cases been associated with exposure to IAP, most frequently related to humidifiers in homes and offices contaminated with bacteria, fungi, or protozoans.

Allergic asthma and extrinsic allergic alveolitis resolve with cessation of exposure to the allergen, but continued exposure in sensitised patients may result in permanent lung damage and death from pulmonary insufficiency.

Humidifier fever is a flu-like illness involving the immune system, in which X-ray abnormalities are usually absent. The exact cause is not clear. The disease may occur among persons exposed to humidification systems contaminated with microbial growth. The symptoms typically occur 4-8 h after the exposure on the first day back at work after a weekend, but resolve within 24 h. Despite continuous exposure the disease does not recur until after the next weekend. Even though pulmonary changes are seen during attacks of humidifier fever, the disease does not lead to permanent lung damage.

## Cancer and Effects on Reproduction associated with Exposure to IAP

Lung cancer is the major cancer which has been associated with exposure to IAP (radon or ETS). Asbestos exposure has been linked to cancer in workers and also in workers' family members, presumably due to asbestos fibres brought into the home on workers' clothing. However, there are no studies associating asbestos exposure in homes or public buildings from asbestos used as a construction material to the development of cancer. Effects on human reproduction have been associated with exposure to chemicals in the environment, but it is as yet unclear to what extent (if any) exposure to IAP is involved.

## Sensory Effects and other Effects on the Nervous System associated with IAP

Sensory effects are defined as the perceptual response to environmental exposures. Sensory perceptions are mediated through the sensory systems and result in a conscious experience of smell, touch, itching, etc. Sensory effects are typically observed in buildings with indoor climate, problems because many chemical compounds found in the indoor air have odorous or mucosal irritation properties. Most indoor air chemicals with a measurable vapour pressure will be odorous when the concentration is high enough.

Sensory effects are important parameters in indoor air quality control for several reasons. They may appear as: (1) adverse health effects on sensory systems (e.g., environmentally-induced sensory dysfunctions); (2) adverse environmental perceptions which may be adverse per se or constitute precursors of disease to come on a long term basis (e.g., annoyance reactions, triggering of hypersensitivity reactions); (3) sensory warnings of exposure to harmful environmental factors (e-g., odour of toxic sulfides, mucosal irritation due to formaldehyde); (4) important tools in sensory bioassays for environmental characterisation (e.g. using the odour criterion for general ventilation requirements or for screening of building materials to find those with low emissions of volatile organic compounds).

The senses responding to environmental exposure are not only hearing, vision, olfaction and taste, but also the skin and mucous membranes. As pointed out by WHO, many different sensory systems that respond to irritants are situated on or near the body surface. Some of these systems tend to respond to an accumulated dose and their reactions are delayed. On the other hand, in the case of odor perception the reaction is immediate but also very much influenced by olfactory fatigue on prolonged exposures.

Responders are often unable to identify a single sensory system as the primary route of sensory irritation by airborne chemical compounds. The sensation of irritation is influenced by a number of factors such as previous exposures, skin temperature, competing sensory stimulation, etc. Since interaction and adaptation processes are characteristic of the sensory systems involved in the perception of odour and mucosal irritation, the duration of exposure influences the perception. Humans integrate different environmental signals to evaluate the total perceived air quality and assess comfort or discomfort. Comfort and discomfort by definition are psychological and for this reason the related symptoms, even when severe cannot be documented without using subjective reports. Sensory effects reported 10 be associated with IAP are in most cases multisensory and the same perceptions or sensations may originate from different sources. It is not known how different sensory perceptions are combined into perceived comfort and into the sensation of air quality. Perceived air quality is for example mainly related to stimulation of both the trigeminus and olfactorius nerves.

Several odorous compounds are also significant mucosal irritants, especially at high concentrations. The olfactory system signals the presence of odorous compounds in the air and has an important role as a warning system. In the absence of instrumentation for chemical detection of small amounts of some odorous vapours, the sense of smell remains the only sensitive indicator system. It is well known that environmental pollution can affect the nervous system. The effects of occupational exposure to organic solvents can be mentioned as an example. A wide spectrum of effects may be of importance, ranging from those at molecular level to behavioural abnormalities. Since the nerve cells of the CNS typically do not regenerate, toxic damage to them is usually irreversible. The nerve cells are highly vulnerable to any depletion in oxygen supply.

## Cardiovascular Effects Associated with IAP

Increased mortality due to Cardiovascular Diseases (CVD) has been associated with exposure to ETS in some groups of non-smoking women married to smokers. Some investigators have also addressed the question whether total mortality is influenced by exposure to ETS, but results have been contradictory. As any effect on mortality would not be expected to occur until after many years of exposure, a problem in these types of study is the accuracy and reliability of the exposure classification. Attempts have also been made to relate ETS to electrocardiographic abnormalities and cardiovascular symptoms, but results have been inconclusive.

Carbon monoxide (CO) exerts its influence primarily through binding to the haemoglobin (Hb) in blood. The affinity of CO to Hb is about 200 times higher than the affinity of oxygen to Hb, so that at relatively low levels of CO in the air. Oxygen is replaced by CO. The percentage of Hb bound to CO

(O/O carboxyhaemoglobin) is a measure of recent exposure to CO. Organs with a high oxygen demand, such as the heart and the brain, are particularly susceptible to a reduced oxygenation caused by CO exposure. Early effects include reduction of time to onset of chest pain in exposed, exercising heart disease patients. At higher levels of exposure, myocardial infarctions may be triggered by CO.

## Basic Control Strategies

There are some basic control methods for lowering concentrations of indoor air pollutants, which are described below:

Source Management includes source removal, source substitution, and source encapsulation. Source management is the most effective control method when it can be practically applied. Source removal is very effective. However, policies and actions that keep potential pollutants from entering indoor are even better than preventing IAQ problems. Source substitution includes actions such as selecting a less toxic art material or interior paint than the products which are currently in use. Source encapsulation involves placing a barrier around the source so that it releases fewer pollutants into the indoor air (e.g., asbestos abatement, pressed wood cabinetry with sealed or laminated surfaces). Local Exhaust is very effecting on removing point sources of pollutants before they can disperse into the indoor air by exhausting the contaminated air outside. Well known examples include restrooms and kitchens where local exhaust is used. Other examples of pollutants that originate at specific points and that can be easily exhausted include science lab and housekeeping storage rooms, printing and duplicating rooms, and vocational/industrial areas such as welding booths. Ventilation through use of cleaner (outdoor) air to dilute the polluted (indoor) air that people are breathing. Generally, local building codes specify the quantity (and sometimes quality) of outdoor air that must be continuously supplied to an occupied area. For situations such as painting, pesticide application, or chemical spills, temporarily increasing the ventilation can be useful in diluting the concentration of noxious fumes in the air. Exposure Control includes adjusting the time of use and location of use. An example of time of use for school students would be to strip and wax floors on Friday after school is dismissed, so that the floor products have a chance to off-gas over the location of use deals with moving the contaminating source as far as possible from occupants, or relocating susceptible occupants. Air Cleaning primarily involves the filtration of particles from the air as the air passes through the ventilation equipment. Gaseous contaminants can also be removed, but in most cases this type of system should be engineered on a case-by-case basis.

# Air Pollutant Concentrations

Air pollutant concentrations, as measured or as calculated by air pollution dispersion modeling, must often be converted or corrected to be expressed as required by the regulations issued by various governmental agencies. Regulations that define and limit the concentration of pollutants in the ambient air or in gaseous emissions to the ambient air are issued by various national and state (or provincial) environmental protection and occupational health and safety agencies.

Such regulations involve a number of different expressions of concentration. Some express the concentrations as ppmv (parts per million by volume) and some express the concentrations as $mg/m^3$ (milligrams per cubic meter), while others require adjusting or correcting the concentrations to reference conditions of moisture content, oxygen content or carbon dioxide content.

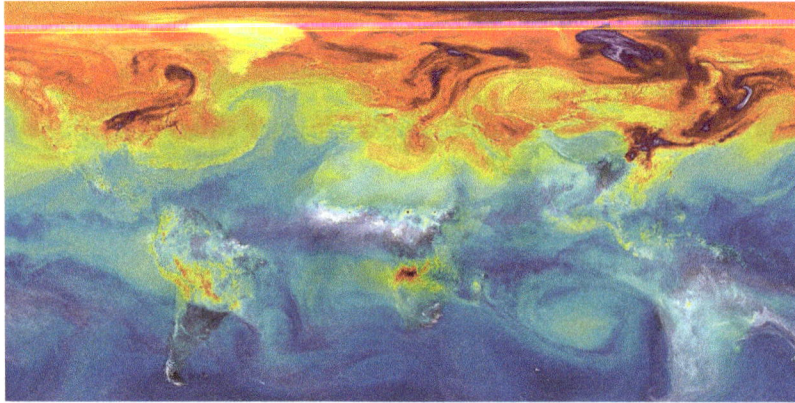

Carbon dioxide in Earth's atmosphere if half of global-warming
emissions are not absorbed.

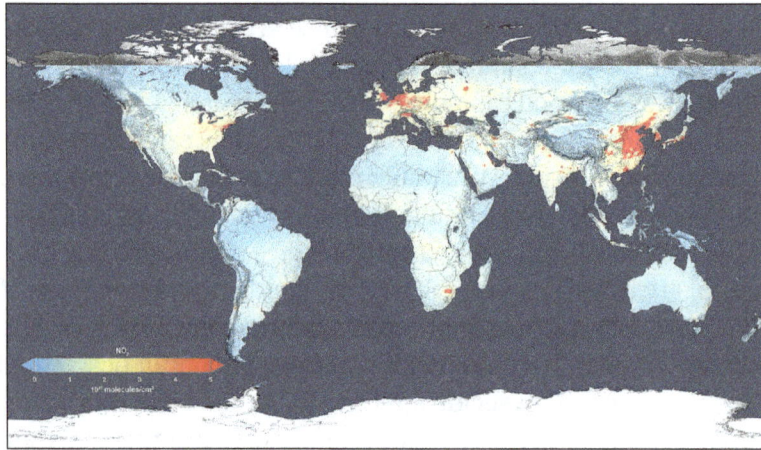

Nitrogen dioxide 2014 - global air quality levels.

## Converting Air Pollutant Concentrations

The conversion equations depend on the temperature at which the conversion is wanted (usually
about 20 to 25 °C). At an ambient sea level atmospheric pressure of 1 atm (101.325 kPa or 1.01325
bar), the general equation is:

$$\text{ppmv} = \text{mg/m}^3 \cdot \frac{(0.082057338 \cdot T)}{M}$$

and for the reverse conversion:

$$\text{mg/m}^3 = \text{ppmv} \cdot \frac{M}{(0.082057338 \cdot T)}$$

where:

- mg/m³ = milligrams of pollutant per cubic meter of air at sea level atmospheric pressure
  and $T$.

- ppmv = air pollutant concentration, in parts per million by volume.

- $T$ = ambient temperature in K = 273. + °C.

- 0.082057338 = Universal gas constant in L atm mol$^{-1}$ K$^{-1}$.

- M = molecular mass (or molecular weight) of the air pollutant.

- 1 atm = absolute pressure of 101.325 kPa or 1.01325 bar.

- mol = gram mole and kmol = 1000 gram moles.

- Pollution regulations in the United States typically reference their pollutant limits to an ambient temperature of 20 to 25 °C as noted above. In most other nations, the reference ambient temperature for pollutant limits may be 0 °C or other values.

- Although ppmv and mg/m³ have been used for the examples in all of the following sections, concentrations such as ppbv (i.e., parts per billion by volume), volume percent, mole percent and many others may also be used for gaseous pollutants.

- Particulate matter (PM) in the atmospheric air or in any other gas cannot be expressed in terms of ppmv, ppbv, volume percent or mole percent. PM is most usually (but not always) expressed as mg/m³ of air or other gas at a specified temperature and pressure.

- For gases, volume percent = mole percent.

- 1 volume percent = 10,000 ppmv (i.e., parts per million by volume) with a million being defined as 10⁶.

- Care must be taken with the concentrations expressed as ppbv to differentiate between the British billion which is 10¹² and the USA billion which is 10⁹ (also referred to as the long scale and short scale billion, respectively).

## Correcting Concentrations for Altitude

Air pollutant concentrations expressed as mass per unit volume of atmospheric air (e.g., mg/m³, µg/m³, etc.) at sea level will decrease with increasing altitude. The concentration decrease is directly proportional to the pressure decrease with increasing altitude. Some governmental regulatory jurisdictions require industrial sources of air pollution to comply with sea level standards corrected for altitude. In other words, industrial air pollution sources located at altitudes well above sea level must comply with significantly more stringent air quality standards than sources located at sea level (since it is more difficult to comply with lower standards). For example, New Mexico's Department of the Environment has a regulation with such a requirement.

The change of atmospheric pressure with altitude (<20 km) can be obtained from this equation:

$$P_h = P \cdot \left(\frac{288 - 6.5h}{288}\right)^{5.2558}$$

Given an air pollutant concentration at sea-level atmospheric pressure, the concentration at higher altitudes can be obtained from this equation:

$$C_h = C \cdot \left(\frac{288 - 6.5h}{288}\right)^{5.2558}$$

where:

  h = altitude, in km.

  $P$ = atmospheric pressure at sea level.

  $P_h$ = atmospheric pressure at altitude h.

  $C$ = air pollutant concentration, in mass per unit volume at sea level atmospheric pressure and specified temperature T.

  $C_h$ = concentration, in mass per unit volume at altitude h and specified temperature T.

As an example, given an air pollutant concentration of 260 mg/m³ at sea level, calculate the equivalent pollutant concentration at an altitude of 2800 meters:

$$C_h = 260 \times [\, \{\, 288 - (6.5)(2.8) \,\} \, / \, 288 \,]^{5.2558} = 260 \times 0.71 = 185 \text{ mg/m}^3.$$

The above equation for the decrease of air pollution concentrations with increasing altitude is applicable only for about the first 10 km of altitude in the troposphere (the lowest atmospheric layer) and is estimated to have a maximum error of about 3 percent. However, 10 km of altitude is sufficient for most purposes involving air pollutant concentrations.

## Correcting Concentrations for Reference Conditions

Many environmental protection agencies have issued regulations that limit the concentration of pollutants in gaseous emissions and define the reference conditions applicable to those concentration limits. For example, such a regulation might limit the concentration of $NO_x$ to 55 ppmv in a dry combustion exhaust gas (at a specified reference temperature and pressure) corrected to 3 volume percent $O_2$ in the dry gas. As another example, a regulation might limit the concentration of total particulate matter to 200 mg/m³ of an emitted gas (at a specified reference temperature and pressure) corrected to a dry basis and further corrected to 12 volume percent $CO_2$ in the dry gas.

Environmental agencies in the USA often use the terms "dscf" or "scfd" to denote a "standard" cubic foot of dry gas. Likewise, they often use the terms "dscm" or "scmd" to denote a "standard" cubic meter of gas. Since there is no universally accepted set of "standard" temperature and pressure, such usage can be and is very confusing. It is strongly recommended that the reference temperature and pressure always be clearly specified when stating gas volumes or gas flow rates.

## Correcting to a Dry Basis

If a gaseous emission sample is analyzed and found to contain water vapor and a pollutant concentration of say 40 ppmv, then 40 ppmv should be designated as the "wet basis" pollutant concentration. The following equation can be used to correct the measured "wet basis" concentration to a "dry basis" concentration:

$$C_{dry\,basis} = \frac{C_{wet\,basis}}{1 - w}$$

where:

    $C$ = concentration of the air pollutant in the emitted gas.

    $w$ = fraction of the emitted exhaust gas, by volume, which is water vapor.

As an example, a wet basis concentration of 40 ppmv in a gas having 10 volume percent water vapor would have a:

    $C_{dry\ basis}$ = 40 ÷ ( 1 - 0.10 ) = 44.4 ppmv.

## Correcting to a Reference Oxygen Content

The following equation can be used to correct a measured pollutant concentration in a dry emitted gas with a measured $O_2$ content to an equivalent pollutant concentration in a dry emitted gas with a specified reference amount of $O_2$:

$$C_r = C_m \cdot \frac{(20.9 - \text{reference volume\%}O_2)}{(20.9 - \text{measured volume\%}O_2)}$$

where:

    $C_r$ = corrected concentration of a dry gas with a specified reference volume % $O_2$.

    $C_m$ = measured concentration in a dry gas having a measured volume % $O_2$.

As an example, a measured $NO_x$ concentration of 45 ppmv in a dry gas having 5 volume % $O_2$ is:

    45 × ( 20.9 - 3 ) ÷ ( 20.9 - 5 ) = 50.7 ppmv of $NO_x$.

when corrected to a dry gas having a specified reference $O_2$ content of 3 volume %.

The measured gas concentration $C_m$ must first be corrected to a dry basis before using the above equation.

## Correcting to a Reference Carbon Dioxide Content

The following equation can be used to correct a measured pollutant concentration in an emitted gas (containing a measured $CO_2$ content) to an equivalent pollutant concentration in an emitted gas containing a specified reference amount of $CO_2$:

$$C_r = C_m \cdot \frac{(\text{reference volume\%}CO_2)}{(\text{measured volume\%}CO_2)}$$

where:

    $C_r$ = corrected concentration of a dry gas having a specified reference volume % $CO_2$.

    $C_m$ = measured concentration of a dry gas having a measured volume % $CO_2$.

As an example, a measured particulates concentration of 200 mg/m³ in a dry gas that has a measured 8 volume % $CO_2$ is:

$$200 \times ( 12 \div 8 ) = 300 \text{ mg/m}^3$$

when corrected to a dry gas having a specified reference $CO_2$ content of 12 volume %.

## Roadway Air Dispersion Modeling

Roadway air dispersion modeling is the study of air pollutant transport from a roadway or other linear emitter. Computer models are required to conduct this analysis, because of the complex variables involved, including vehicle emissions, vehicle speed, meteorology, and terrain geometry. Line source dispersion has been studied since at least the 1960s, when the regulatory framework in the United States began requiring quantitative analysis of the air pollution consequences of major roadway and airport projects. By the early 1970s this subset of atmospheric dispersion models were being applied to real world cases of highway planning, even including some controversial court cases.

### How the Model Works?

The basic concept of the roadway air dispersion model is to calculate air pollutant levels in the vicinity of a highway or arterial roadway by considering them as line sources. The model takes into account source characteristics such as traffic volume, vehicle speeds, truck mix, and fleet emission controls; in addition, the roadway geometry, surrounding terrain and local meteorology are addressed. For example, many air quality standards require that certain near worst case meteorological conditions be applied.

The calculations are sufficiently complex that a computer model is essential to arrive at authoritative results, although workbook type manuals have been developed as screening techniques. In some cases where results must be refereed (such as legal cases), model validation may be needed with field test data in the local setting; this step is not usually warranted, because the best models have been extensively validated over a wide spectrum of input data variables.

The product of the calculations is usually a set of isopleths or mapped contour lines either in plan view or cross sectional view. Typically these might be stated as concentrations of carbon monoxide, total reactive hydrocarbons, oxides of nitrogen, particulate or benzene. The air quality scientist can run the model successively to study techniques of reducing adverse air pollutant concentrations (for example, by redesigning roadway geometry, altering speed controls or limiting certain types of trucks). The model is frequently utilized in an Environmental Impact Statement involving a major new roadway or land use change which will induce new vehicular traffic.

### The Theory

The resulting solution for an infinite line source is:

$$x = \int_0^\infty \frac{q}{\pi \left(ucdx^2\right)\left(cos\alpha\right)} \left(\exp\frac{y^2}{2c^2x^2}\right) dx$$

where:

$x$ is the distance from the observer to the roadway.

$y$ is the height of the observer.

$u$ is the mean wind speed.

$\alpha$ is the angle of tilt of the line source relative to the reference frame.

$c$ and $d$ are the standard deviation of horizontal and vertical wind directions (measured in radians) respectively.

This equation was integrated into a closed form solution using the error function (erf), and variations in geometry can be performed to include the full infinite line, line segment, elevated line, or arc made from segments. In any case one can calculate three-dimensional contours of resulting air pollutant concentrations and use the mathematical model to study alternative roadway designs, various assumptions of worst case meteorology or varying traffic conditions (for example, variations in truck mix, fleet emission controls, or vehicle speed).

The ESL research group also extended their model by introducing the area source concept of a vertical strip to simulate the mixing zone on the highway produced by vehicle turbulence. This model too was validated in 1971 and showed good correlation with field test data.

## Example Applications of the Model

There were several early applications of the model in somewhat dramatic cases. In 1971 the Arlington Coalition on Transportation (ACT) was the plaintiff in an action against the Virginia Highway Commission over the extension of Interstate 66 through Arlington, Virginia, having filed a suit in the federal district court. The ESL model was used to produce calculations of air quality in the vicinity of the proposed highway. ACT won this case after a decision by the U.S. Fourth Circuit Court of Appeals. The court paid special attention to the plaintiff's expert calculations and testimony projecting that air quality levels would violate Federal ambient air quality standards as set forth in the Clean Air Act.

Roadway air dispersion modeling is also done for curved roadways-North-South Express Highway.

A second contentious case took place in East Brunswick, New Jersey where the New Jersey Turnpike Authority planned a major widening of the Turnpike. Again the roadway air dispersion model was employed to predict levels of air pollution for residences, schools and parks near the Turnpike. After an initial hearing in Superior Court where the ESL model results were set forth, the judge ordered the Turnpike Authority to negotiate with the plaintiff, Concerned Citizens of East Brunswick and develop air quality mitigation for the adverse effects. The Turnpike Authority hired ERT as its expert, and the two research teams negotiated a settlement to this case using the newly created roadway air dispersion models.

## More Recent Model Refinements

The CALINE3 model is a steady-state Gaussian dispersion model designed to determine air pollution concentrations at receptor locations downwind of highways located in relatively uncomplicated terrain. CALINE3 is incorporated into the more elaborate CAL3QHC and CAL3QHCR models. CALINE3 is in widespread use due to its user friendly nature and promotion in governmental circles, but it falls short of analyzing the complexity of cases addressed by the original Hogan-Venti model. CAL3QHC and CAL3QHCR models are available in the Fortran programming language. They have options to model either particulate matter or carbon monoxide, and include algorithms to simulate queued traffic at signalized intersections.

In addition, several more recent models have been developed that employ non-steady state Lagrangian puff algorithms. The HYROAD dispersion model has been developed through the National Cooperative Highway Research Program's Project 25-06, incorporating ROADWAY-2 model puff and steady-state plume algorithms.

The TRAQSIM model was developed in 2004 as part of a Ph.D dissertation with support by the U.S. Department of Transportation's Volpe National Transportation Systems Center Air Quality Facility. The model incorporates dynamic vehicle behavior with a non-steady state Gaussian puff algorithm. Unlike HYROAD, TRAQSIM combines traffic simulation, second-by-second modal emissions, and Gaussian puff dispersion into a fully integrated system (a true simulation) that models individual vehicles as discrete moving sources. TRAQSIM was developed as a next generation model to be the successor to the current CALINE3 and CAL3QHC regulatory models. The next step in the development of TRAQSIM is to incorporate methods to model the dispersion of particulate matter (PM) and hazardous air pollutants (HAPs).

Several models have been developed that handle complex urban meteorology resulting from urban canyons and highway configurations. The earliest such model development was by the Air Pollution Control Office of the U.S. EPA in conjunction with New York City. The model was successfully applied to the Spadina Expressway in Toronto by Jack Fensterstock of the New York City Department of Air Resources,. Other examples include the Turner-Fairbank Highway Research Center's Canyon Plume Box model, now in version 3 (CPB-3), the National Environmental Research Institute of Denmark's Operational Street Pollution Model (OSPM), and the MICRO-CALGRID model, which includes photochemistry, allowing for both primary and secondary species to be modeled. Cornell University's CTAG model, which resolves vehicle-induced turbulence (VIT), road-induced turbulence (RIT), chemical transformation and aerosol dynamics of air pollutants using turbulence reacting flow models. The CTAG model has also been applied to characterize highway-building environments and study effects of vegetation barriers on near-road air pollution.

# Air Pollution Dispersion Model

Air quality models are used to predict ground level concentrations down point of sources. The object of a model is to relate mathematically the effects of source emissions on ground level concentrations, and to establish that permissible levels are, or are not, being exceeded. Models have been developed to meet these objectives for a variety of pollutants and time circumstances.

Models may be described according to the chemical reactions involved. So-called nonreactive models are applied to pollutants such as CO and $SO_2$ because of the simple manner in which their chemical reactions can be represented. Reactive models address complex multiple-species chemical mechanism common to atmospheric photochemistry and apply to pollutants such as NO, $NO_2$, and $O_3$.

Models can be described as simple or advanced based on the assumptions used and the degree of sophisticated with which the important variables are treated. Advanced models have been developed for such problems as photochemical pollution, dispersion in complex terrain, long-range transport, and point sources over flat terrain. The most widely used models for predicting the impact of relative unreactive gases, such as $SO_2$, released from smokestacks are based on Gaussian diffusion.

In Gaussian models, the spread of a plume in vertical horizontal directions is assumed to occur by simple diffusion along the direction of the mean wind. The maximum ground level concentration is calculated by means of the following equation:

$$C_x = \frac{Q}{\pi \sigma_y \sigma_z u} e^{-1/2 \left[\frac{H}{\sigma_z}\right]^2} e^{-1/2 \left[\frac{y}{\sigma_y}\right]^2}$$

where:

Cx  =  ground level concentration at some distance x downwind (g/m³).

Q   =  average emission rate (g/sec).

u   =  mean wind speed (m/sec).

H   –  effective stack height (m).

$\sigma_y$  =  standard deviation of wind direction in the horizontal (m).

$\sigma_z$  =  standard deviation of wind direction in the vertical (m).

y   =  off-centerline distance (m).

e   =  natural log equal to 2.71828.

The parameters $\sigma_y$ and $\sigma_z$ describe horizontal and vertical dispersion characteristics of a plume at various distances downwind of a source as function of different atmospheric stability conditions. Values are determined from the graphs found in the figure.

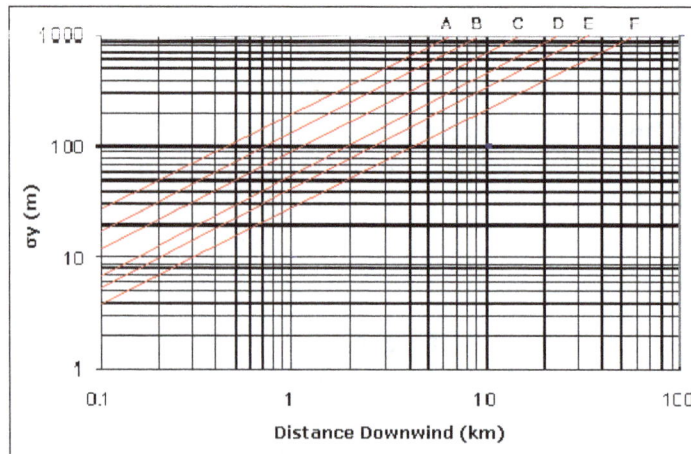

Horizontal Dispersion Coefficent.

The effective stack height H is equal to the physical stack height (h) plus the height of the plume (plume rises, Δh) determined from where the plume bends over. Plume rises must be calculated from model equations before the effective stack height can be calculated.

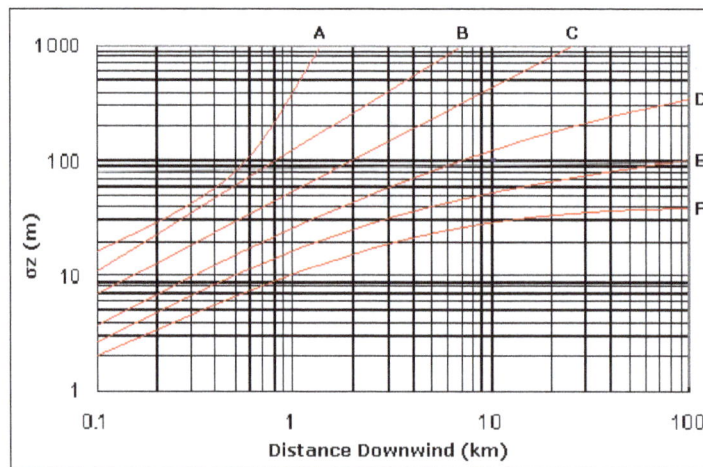

Vertical Dispersion Coefficent.

For purposes of illustration, let us determine the ground level concentration $(C_x)$ at some downwind distance (x). For the following conditions let us calculate the ground level concentrations at 10 km directly downwind.

A power plant burning 9 tons of 2.5% sulfur coal/hr emits $SO_2$ at a rate of 113 g/sec. The effective stack height is 100 m, and the wind speed is 3 m/sec. It is 1 hour before sunrise, and the sky is clear. Since the off centerline distance (Y) in this case is equal to O, the following equation reduces to:

$$C_x = \frac{Q}{\pi \sigma_y \sigma_z u} e^{-1/2} \left[ \frac{H}{\sigma_z} \right]^2$$

Therefore:

$$C_x = 44 \mu g / m^3$$

Table: Key to stability classes.

| Wind speed 10m | Day | | | Night | |
|---|---|---|---|---|---|
| (m/sec) | Incoming solar radiation | | | Thinly Overcast | |
| | Strong | Moderate | Slight | >4/8 Cloud | <3/8Cloud |
| <2 | A | A-B | B | E | F |
| 3-Feb | A-B | B | C | D | E |
| 5-Mar | B | B-C | C | D | D |
| <6 | C | D | D | D | D |

From table, the atmospheric stability classes for the condition described is F. It represent a night-time condition with <37.5% cloud cover. The horizontal dispersion coefficient $\sigma_y$ for a downtime distance of 5 km for atmospheric stability class F is approximately 90 m the vertical dispersion coefficient $\sigma_z$ is approximately 20 m.

The ground level concentration of $SO_2$ from this source would be approximately 44 µg/m³ under the conditions given.

Although the use of air quality models is the subject of considerable controversy, there's a general agreement that there a few alternatives to the use of models, particulately to make decisions on an action which is know in advance to pose potential environmental problem. The debate arises as to which models should be used, and the interpretation of models results. The underlying question such in debates is how well, or how accurately, does the model predict concentrations under the specific circumstances, since model accuracy may vary from 30% to a factor of 2 or more? If a model is conservative, i.e., it over-predicts ground level concentrations, a source may be required to install costly control equipment unnecessarily. Less conservative models may under-predict concentrations and thus violations of air quality standards may occur. The uncertainty associated with input variables, such as wind data, and source emission data. Such data are usually estimated and not well documented.

Air pollution may be caused by various processes, either natural or anthropogenic (man-made). Some of them leave evident traces in the air; others can go unnoticed unless specific tests are conducted - or until you become ill from their effects.

## Natural Causes

- Volcanic activities – Volcanic eruptions emit a series of toxic gases (including sulfur and chlorine) as well as particulate matter (ash particles) but are usually restricted to localized areas.

- Winds and air currents – Can mobilize pollutants from the ground and transport them over large areas.

- Wildfires – Add carbon monoxide, as well as particulate matter, to the atmosphere (containing organic contaminants such as PAHs); could affect significant areas, although in general they are restricted and may be contained.

- Microbial decaying processes – Microorganisms which are present in any environment have a major role in natural decaying processes of living organisms as well as environmental contaminants; this activity results in the natural release of gases especially methane gas.

- Radioactive decay processes – For example, radon gas is emitted due to natural decay processes of Earth's crust which has potential to accumulate in enclosed spaces such as basements.

- Increasing temperatures – Contribute to an increase in the amounts of contaminants volatilizing from polluted soil and water into the air.

## Anthropogenic Causes

- Mining and smelting – Emit into the air a variety of metals adsorbed on particulate matter that is suspended in the air due to crushing & processing of mineralogical deposits.

- Mine tailing disposal – Due to their fine particulate nature (resulting after crushing and processing mineral ores) constitute a source of metals to ambient air which could be spread by the wind over large areas.

- Foundry activities – Emit into the air a variety of metals absorbed on particulate matter that is suspended in the air due to processing of metallic raw materials (including the use of furnaces).

- Various industrial processes may emit both organic and inorganic contaminants through accidental spills and leaks of stored chemicals or the handling and storage of chemicals – especially of volatile inorganic chemicals.

- Transportation – Emits a series of air pollutants (gases – including carbon monoxide, sulfur oxides, and nitrogen oxides - and particulate matter) through the tailpipe gases due to internal combustion of various fuels (usually gasses such as oxides of carbons, of sulfur, of nitrogen, as well as organic chemicals as PAHs)

- Construction and Demolition activities – Pollute the air with various construction materials. Of special threat is the demolition of old buildings which may contain a series of banned chemicals such as PCBs, PBDEs, asbestos.

- Coal Power Plants – When burning coal this may emit a series of gases as well as particulate matter with metals (such as As, Pb, Hg) and organic compounds (especially PAHs).

- Heating of buildings – Emits a series of gases and particulate matters due to burning fossil fuels.

- Waste Incineration – Depending on waste composition, various toxic gases, and particulate matter is emitted into the atmosphere.

- Landfill disposal practices – Usually generate methane due to the intensification of natural microbial decaying activity in the disposal area.

- Agriculture – Pollute the air usually through emissions of ammonia gas and the application of pesticides/herbicides/insecticides which contain toxic volatile organic compounds.

- Control burning in forest and agriculture management – Includes controlled burning that will emit gases and particulate matter.

- Military activities – May introduce toxic gases through practices and training.

- Smoking – Emits a series of toxic chemicals including a series of organic and inorganic chemicals, some of which are carcinogenic.

- Storage and use of household products such as paint, sprays, varnish, etc that contains organic solvents which volatilize in the air (hence the smell we all feel while using them).

- Dry cleaned clothes - May retain and emit in the atmosphere small amounts of chlorinated solvents (such as PCE) or petroleum solvents that have been used by the dry cleaners; this could eventually create a health risk if the clothes returned from the dry cleaners are stored in enclosed indoor spaces.

# Mobile Source Air Pollution

Mobile source air pollution includes any air pollution emitted by motor vehicles, airplanes, locomotives, and other engines and equipment that can be moved from one location to another. Many of these pollutants contribute to environmental degradation and have negative effects on human health. To prevent unnecessary damage to human health and the environment, environmental regulatory agencies such as the U.S. Environmental Protection Agency have established policies to minimize air pollution from mobile sources. Similar agencies exist at the state level. Due to the large number of mobile sources of air pollution, and their ability to move from one location to another, mobile sources are regulated differently from stationary sources, such as power plants. Instead of monitoring individual emitters, such as an individual vehicle, mobile sources are often regulated more broadly through design and fuel standards. Examples of this include corporate average fuel economy standards and laws that ban leaded gasoline in the United States. The increase in the number of motor vehicles driven in the U.S. has made efforts to limit mobile source pollution challenging. As a result, there have been a number of different regulatory instruments implemented to reach the desired emissions goals.

Cars are Major Sources of Mobile Air Pollution.

## Broad Classification

Airplanes Produce Significant Levels of Pollution Emissions.

There are a number of different mobile sources of air pollution, some contributing more to pollution than others. As mentioned previously, mobile sources are regulated differently from stationary sources due to the large number of sources and their ability to move from one location to another. Different mobile sources operate differently and generate different emission types and levels. The E.P.A. differentiates between mobile sources by classifying them as either on-road vehicles or non-road vehicles. On-road vehicles and non-road vehicles are often subject to different regulations.

### Road Sources

- Cars

- Light Duty and Heavy Duty Trucks

- Buses

- Motorbikes

### Non-road sources

- Aircraft

- Motorboats (Diesel and Gasoline)

- Locomotives

- Construction Equipment

### Major Regulated Mobile Source Pollutants

There are a number of different pollutants that are emitted by mobile sources. Some make up a large portion of the total air concentration for that particular pollutant while others do not make up as much of the total air concentration.

- Carbon Monoxide: Carbon monoxide forms when carbon in fuel does not burn completely (incomplete combustion). The main source of carbon monoxide in air is vehicle emissions. As much as 95 percent of the carbon monoxide in typical U.S. cities comes from mobile sources, according to EPA studies. Carbon monoxide is harmful because it reduces oxygen delivery to the body's organs and tissues. It is most harmful to those who suffer from heart and respiratory disease.

- Carbon Dioxide: Carbon dioxide is one of the most prominent greenhouse gasses emitted by motor vehicles. In 2006, 23.6% of the total inventory of U.S. greenhouse gasses were derived from motor vehicles. The compound is generated as a byproduct of the combustion of any fuel source containing carbon.

- Nitrogen Oxides: Nitrogen oxides form when fuel burns at high temperatures, such as in motor vehicle engines. Mobile sources are responsible for more than half of all nitrogen oxide emissions in the United States. Both on-road and non-road mobile sources are major nitrogen oxide polluters. These problems include ozone and smog.

- Hydrocarbons: Hydrocarbons are a precursor to ground-level ozone, a serious air pollutant in cities across the United States. A key component of smog, ground-level ozone is formed by reactions involving hydrocarbons and nitrogen oxides in the presence of sunlight. Hydrocarbon emissions result from incomplete fuel combustion and from fuel evaporation. Ground-level ozone causes health problems such as difficulty breathing, lung damage, and reduced cardiovascular functioning.

- Particulate Matter: Atmospheric particulate matter or airborne particulate matter is the term for solid or liquid particles found in the air. Some particles are large or dark enough to be seen as soot or smoke, but fine particulate matter is tiny and is generally not visible to the naked eye. Fine particulate matter is a health concern because very fine particles can reach the deepest regions of the lungs. Health effects include asthma, difficult or painful breathing, and chronic bronchitis, especially in children and the elderly.

- Air Toxics: The EPA lists over 1100 individual compounds which are classified as air toxics. These compounds are emitted by mobile sources, mostly due to the chemical nature of the fuel source. These compounds are known or expected to cause serious physical damages including cancer, reproductive, and developmental side effects. The comprehensive list of regulated air toxics can be found at the EPA's website. EPA - Mobile Source Air Toxics.

## U.S. Enforcement Agencies

## Federal Agencies

- Environmental Protection Agency: The Environmental Protection Agency's Office of Air and Radiation (OAR) develops national programs, policies, and regulations for controlling air pollution and radiation exposure. OAR is concerned with pollution prevention and energy efficiency, indoor and outdoor air quality, industrial air pollution, pollution from vehicles and engines, radon, acid rain, stratospheric ozone depletion, climate change, and radiation protection.

- Department of Energy: The Department of Energy's clean air compliance activities are overseen by its Office of Health, Safety, and Security.

- Department of Transportation:

  o Federal Aviation Administration: Practically all aviation emission sources are independently regulated through equipment specific regulations, standards and recommended practices, and operational guidelines, which are established by a variety of organizations. For example, on-road vehicles, which take passengers to and from the airport, meet stringent Federal tailpipe standards set by EPA. Stationary sources on the airport, like power boilers and refrigeration chillers, must meet independent state regulations. And FAA certification is required for essentially all aviation equipment and processes. For example there are more than 60 standards that apply to aircraft engine design, materials of construction, durability, instrumentation and control, and safety, among others. These are in addition to the Fuel Venting and Exhaust Emission Requirements for Turbine Engine Powered Airplanes, which guide compliance with EPA's aircraft exhaust emission standards. The International Civil Aviation Organization (ICAO) is a United Nations intergovernmental body responsible for worldwide planning, implementation, and coordination of civil aviation. ICAO sets emission standards for jet engines. These are the basis of FAA's aircraft engine performance certification standards, established through EPA regulations.

  o Federal Highway Administration: The FHWA, EPA, the Health Effects Institute, and others have funded and conducted research studies to try to more clearly define potential risks from mobile source air toxics emissions associated with highway projects. The FHWA policies and procedures for implementing NEPA is prescribed by regulation in 23 CFR 771.

  o National Highway Traffic Safety Administration: NHTSA administers the CAFE program, and the Environmental Protection Agency (EPA) provides the fuel economy data. NHTSA sets fuel economy standards for cars and light trucks sold in the U.S. while EPA calculates the average fuel economy for each manufacturer.

## State-level Agencies

EPA has ten regional offices, each of which is responsible for the execution of programs within several states and territories. EPA's website provides a detailed list of state agencies which administer the environmental regulations at the state level. California is the only state which has its own regulatory agency, the California Air Resources Board (CARB). The other states are allowed to follow CARB or federal regulations.

## Enforcement Mechanisms and Policy Instruments

Federal, state, and local governments utilize a wide range of policy instruments to control pollution from mobile sources. On the federal level, many different agencies are responsible for regulating, or at least creating policies to limit, pollution from mobile sources. This is necessary given the broad range of objects that are considered "mobile sources," from aircraft and off-road vehicles, to locomotives and on-road vehicles. The Federal Aviation Administration, for example, establishes standards to limit emissions from aircraft, whereas the U.S. Department of Transportation and Environmental Protection Agency administer various aspects of on-road vehicle fuel economy regulations. On the state level, mandatory vehicle emissions-testing programs are often required as part of the annual motor-vehicle registration process.

## Labeling Policies

Many governments throughout the world require manufacturers of particular products to attach information-related labels to their products. Common examples in the United States include food nutrition and ingredient labels for food products, Surgeon General labels on alcohol and tobacco products, and labels for common household pesticides. Like mobile sources of air pollution, there is a broad range of products that may require government labeling regulation, therefore numerous federal agencies oversee various label-related regulation programs. For example, the US Food and Drug Administration oversees food nutrition and ingredient label regulations, whereas the US Environmental Protection Agency sets specific standards for the labeling of pesticides.

The primary goal of labeling regulations is to provide consumers and other product users with important information about the product. Essentially, labeling policies are designed to correct the market failure of imperfect information. For consumers to make the best decisions when allocating scarce resources, such as income, detailed information about particular products may be required. In this sense, labels also help correct information asymmetries that often exist within many market transactions.

| EPA DOT | Fuel Economy and Environmental Comparison |
| --- | --- |

**A+**

Smartphone

The above grade reflects fuel economy and greenhouse gases. Grading system ranges from A+ to D.

**website.here**

Over five years, this vehicle

saves **$6,900** in fuel costs compared to the average vehicle.

Electric Vehicle

| Range (miles) | kW-hrs/ 100 Miles | MPGe City | MPGe Highway | CO₂ g/mile (tailpipe only) | Annual fuel cost |
| --- | --- | --- | --- | --- | --- |
| 99 | 34 | 102 | 94 | 0 | $618 |

Combined MPGe    CO₂ g/mile    Other Air Pollutants

- Fuel economy for all midsize cars ranges from 12 to 103 MPGequivalent. MPGequivalent: 33.7 kW-hrs = 1 gallon gasoline energy.
- Annual fuel cost based on 15,000 miles per year at 12 cents per kW-hr.

Visit website.here to calculate estimates personalized for your driving, and to download the Fuel Economy Guide (also available at dealers).

In the United States, all new cars and light-duty trucks are required to have labels that display specific fuel economy information. The US Environmental Protection Agency calculates the average fuel economy for each vehicle manufacturer, and provides the data to the National Highway Traffic Safety Administration (NHTSA), which administers and enforces the Corporate Average Fuel Economy (CAFE) program. The purpose of the program is (1) to reduce emissions by requiring vehicle manufacturers to meet minimum fuel economy levels, and (2) to provide consumers with fuel economy information before purchasing new vehicles.

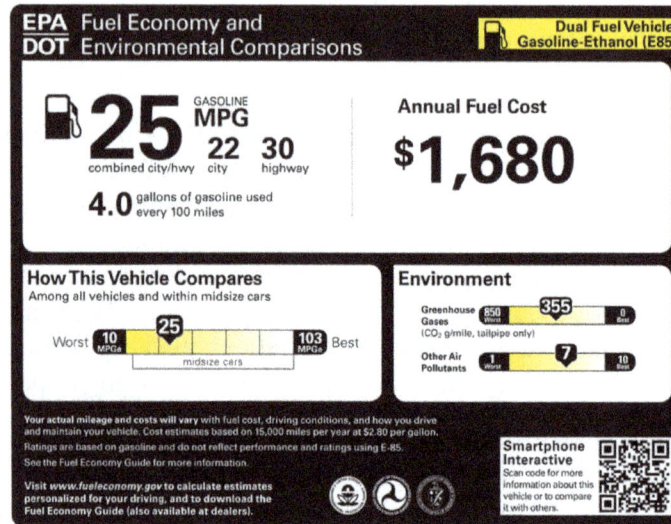

Proposed CAFE Label.

EPA and NHTSA are redesigning the labels to provide even more information to consumers. The new labels will, for the first time, provide information about each vehicle's greenhouse gas emissions, as required by the Energy Independence and Security Act of 2007. The agencies are proposing two different label designs and are seeking public comments about which labels will be most helpful to consumers.

## Taxes

Another common policy instrument used by governments to influence market behavior is taxation. In the case of mobile source air pollution, the United States government has established many different taxes to limit emissions from various mobile sources. Perhaps one of the most well known is the gas guzzler tax, established by the Energy Tax Act of 1978. The act set minimum fuel economy standards for all new cars sold in the United States.

The tax is levied against manufacturers of new *cars* that fail to meet the minimum fuel economy level of 22.5 miles per gallon. The tax does not apply to minivans, sport utility vehicles, or pickup trucks, as these made up a small portion of the US fleet when the tax was established in 1978. Manufacturers pay a level of tax based upon the average fuel economy for each particular vehicle produced, ranging from $1,000 for vehicles achieving at least 21.5 but less than 22.5 MPG, to $7,000 for each vehicle achieving less than 12.5 MPG. Vehicles that achieve a minimum average fuel economy of 22.5 MPG are not subject to the gas guzzler tax.

## Tax Credits

Governments may also offer tax credits to encourage certain types of behavior within market economies. For example, if a government wants to encourage consumers to purchase more fuel-efficient vehicles, the government could offer tax credits to effectively lower the price of each vehicle. The logic of this approach is consistent with the laws of supply and demand, namely, that as the price of a good decreases, the quantity demanded of that good will increase. This is true given that other important factors, such as current levels of supply and demand, remain constant.

The US federal government currently utilizes numerous tax credits to reduce emissions from mobile sources. One of the more common tax credits is the "Qualified Plug-In Electric Drive Motor Vehicle Tax Credit." This credit is available "for the purchase of a new qualified plug-in electric drive motor vehicle that draws propulsion using a traction battery that has at least four kilowatt hours of capacity, uses an external source of energy to recharge the battery, has a gross vehicle weight rating of up to 14,000 pounds, and meets specified emission standards." The credit ranges from \$2,500 to \$7,000, depending upon the vehicle's weight rating. Consumers who purchase the new Chevrolet Volt are eligible for the full \$7,500 credit. Another tax credit targeted at consumers is the "Fuel Cell Motor Vehicle Tax Credit," which was originally set at \$8,000 for the purchase of qualified light-duty fuel cell vehicles. On December 31, 2009, the tax credit was reduced to \$4,000.

Tax credits to limit mobile source pollution can also be targeted at producers of particular products. For example, "Advanced Biofuel Production Payments" are available to "eligible producers of advanced biofuels," or for fuels derived from "renewable biomass other than corn kernel starch." Such producers "may receive payments to support expanded production of advanced biofuels," dependent upon the "quantity and duration of production by the eligible producer; the net nonrenewable energy content of the advanced biofuel, if sufficient data is available; the number of producers participating in the program; and the amount of funds available." While many critics have argued that biofuels can actually increase greenhouse gas emissions, research from the US Department of Energy indicates that biofuels "burn cleaner than gasoline, resulting in fewer greenhouse gas emissions, and are fully biodegradable, unlike some fuel additives."

## Voluntary Programs

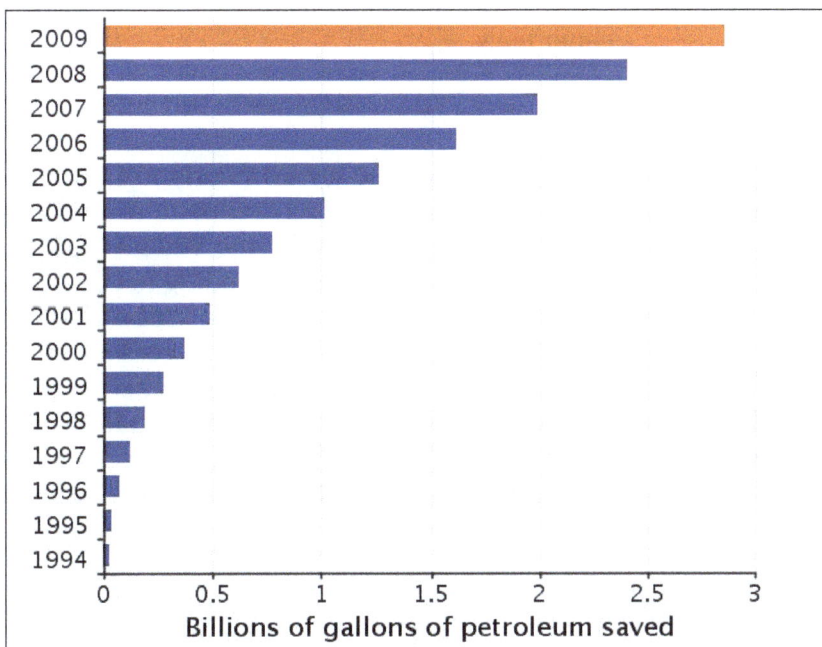

The US Department of Energy's "Clean Cities" program has saved more than 3 billion US gallons (11,000,000 m³) of petroleum since its inception in 1993.

Other important policy instruments that can be utilized by governments are voluntary programs. These programs bring together various stakeholders with the goal of achieving some particular policy outcome. The Department of Energy, for example, created the "Clean Cities" program to reduce petroleum use in the transportation sector. The Clean Cities program partners with more than 80 volunteer organizations throughout the United States, developing public-private partnerships that promote alternative fuels and advanced vehicles, fuel blends, fuel economy, hybrid vehicles, and idle reduction. The three primary goals of the program are:

1.  Replacement: Replace petroleum used in the transportation sector with alternative and renewable fuels.

2.  Reduction: Reduce petroleum use by promoting smarter driving practices, idle reduction, fuel-efficient vehicles, and advanced technologies.

3.  Elimination: Eliminating petroleum use by encouraging greater use of mass transit systems, trip-elimination measures, and congestion mitigation.

The program was initiated in 1993 and has saved nearly 3 billion US gallons (11,000,000 m³) of petroleum since its inception.

Another example of a voluntary program is the Environmental Protection Agency's "SmartWay Transport Partnership." This voluntary partnership between the EPA and the ground freight industry is designed to reduce greenhouse gases and air pollution through increased fuel efficiency programs. EPA provides partners with "benefits and services that include fleet management tools, technical support, information, public recognition, and use of the SmartWay Transport Partner logo."

"Clean Construction USA" is an additional voluntary program administered by EPA that promotes the reduction of diesel exhaust emissions from construction equipment and other construction vehicles. The program encourages proper operations and maintenance, the use of emission-reducing technologies, and the use of cleaner fuels.

## Subsidies

Subsidies are another powerful policy tool used by governments to influence economic behavior. Subsidies can take many forms, ranging from tax credits to direct cash payments. To limit mobile source pollution from airports, for example, the Federal Aviation Administration's "Voluntary Airport Low Emission Program" provides funding to U.S. commercial service airports located in air quality non attainment and maintenance areas. While the funding can be used to reduce emissions from both mobile and stationary sources at the airport, much of the program's emphasis is on mobile source emission reduction. The program promotes the use of electric ground support equipment, such as electric bag tugs that take luggage from the airplane to the baggage claim. Other airport equipment that can be electronically operated include various types of belt loaders, along with the pushback tractors that assist airplanes when departing from the gate.

Another important goal of the program is to install underground fuel hydrants at airports. These would eliminate the need for fuel trucks, an important source of mobile emissions. The Voluntary Airport Low Emission Program was established under the Vision 100 Century of Aviation Reauthorization Act of 2003.

## Command and Control: Performance Standards

Numerous states have emissions-testing programs to limit pollution from on-road vehicles, such as cars and light-duty trucks. Each of these vehicles must meet specific emissions targets before being allowed to obtain or renew vehicle registrations. Many of these programs are administered on the local and county level. For example, the Clean Air Car Check is a vehicle emissions-testing program for all vehicles registered in Lake and Porter counties in Indiana. The two counties were designated as non-attainment areas for ozone levels in 1977 by the Environmental Protection Agency. By 1990, the two counties were reclassified as severe non-attainment areas, a designation which requires states to create State Implementation Plans to attain and maintain certain air pollution standards. Although the counties were again reclassified in 2010, this time as attainment areas, the two counties will maintain their vehicle inspection and maintenance program because it is a "key piece of Indiana's plan to prevent backsliding so that the area can remain in attainment."

## Corporate Average Fuel Efficiency Standard

According to the Corporate Average Fuel Economy standard (CAFE) regulation, which was enacted in 1975, every seller of automobiles in the US had to achieve by 1985 a minimum sales-weighted average fuel efficiency of 27.5 miles per gallon (MPG). This standard had to be achieved for domestically produced and imported cars separately. Failure to meet the prescribed standard incurred a penalty of $5 per car per 1/10 of a gallon that the corporate average fuel economy fell below the standard. The first idea about the environmental impact of the CAFE regulation can be obtained by examining its effects on the average fuel efficiency of domestic and foreign firms; these effects are largest for the domestic production of US manufacturers, whose corporate average fuel efficiency would be lower by 1.2 MPG in the absence of CAFE standards. CAFE standards also lead to approximately 19 million US gallons (72,000 m³) fuel consumption savings per year. Contrary to the CAFE standards, gasoline taxes affect not only new but also used cars, so that there is no reason to expect any substitution towards less fuel efficient used cars when taxes are raised. Small tax increases are insufficient to induce fuel cost savings of the same order of magnitude as CAFE.

## Marketable Allowances

## Leaded Gasoline

Lead was originally added to fuel as an additive to prevent engine knocking. In the 1970s, virtually all gasoline used in the United States contained lead with an average concentration of almost 2.4 grams per gallon. By the mid 1970s, the EPA began formulating plans to phase lead out of fuel for two main reasons. There was growing concern over lead's potential effects on human health, especially with respected hypertension and cognitive development in children. Additionally, the introduction of the catalytic converter in new automobiles manufactured after 1975 required an adjustment to the fuel standards. Catalytic converters were utilized in new automobiles to help meet the hydrocarbon, carbon monoxide, and nitrogen oxide emission standards mandated by the 1970 Clean Air Act. Unfortunately, the catalytic converters could only function properly with unleaded fuel.

In order to protect human health and ensure that catalytic converters were operating properly, the EPA required that the average lead content of all gasoline sold be reduced from 1.7 grams per gallon after January 1, 1975 to 0.5 grams per gallon by January 1, 1979. Eventually, the

EPA lowered the average lead concentration standard goal to 0.1 gm/gal by January 1, 1986. The EPA defined "averages" in a way that allowed refiners who owned more than one refinery to average or "trade" among refineries to satisfy their lead limits each quarter. Taking note of the trading that was taking place, the EPA permitted refiners to bank credits for use until the end of 1987. EPA enforcement relied on reporting requirements and random testing of gasoline samples.

The EPA has officially concluded its effort to phase out lead in fuel. As of 1996, manufacturers are no longer required to place "unleaded fuel only" labels on the dashboard and on or around the fuel filler inlet area of each new motor vehicle. Additionally, several record keeping and reporting requirements for gasoline refiners and importers have been lifted. Critics have viewed the lead credit trading program as a successful implementation of a cap and trade system allowing for the gradual reduction of a pollutant. Lead credit trading as a percentage of lead use rose above 40 percent by 1987. An estimated 20 percent of refineries participated in trading early in the program, eventually rising to 60 percent of refineries.

### Benzene in Gasoline

In 2007, the Mobile Source Air Toxics Rule was created to help limit the hazardous emissions generated as a result of fuel combustion in mobile sources. Benzene is one particular component of gasoline that is known to pose a hazard to human health. In 2007, benzene concentrations in gasoline averaged 1% by volume. The EPA mandated refiners and importers to begin producing gasoline with annual an average benzene content no greater than 0.62% beginning in 2011. The EPA has listed certain technologies that can be utilized in order to achieve the new standards, but refiners can petition the EPA to approve additional technologies.

Refiners and importers could earn credits by reducing benzene levels below 0.62% before 2011. These credits could be auctioned to other companies, essentially creating a marketable allowance approach for reducing benzene content in gasoline. The nationwide banking and trading system does nave some limitations. No individual refiner or importer could produce gasoline with benzene concentrations exceeding 1.3% by volume, even with credits.

## Marine Shipping

While essential to the world's economy and well-being, the commercial marine shipping industry is a major contributor to global air pollution and without action, the industry's emissions are expected to increase.

These emissions can harm human health and our environment. New regulations and practical initiatives are planned or in force to reduce the amount of air pollution produced by ships.

## Marine Shipping's Impact on World Air Pollution

Ships move approximately 80% of the world's goods. When compared to other forms of transportation, marine shipping is the most energy-efficient way to move large volumes of cargo.

Like all other forms of transportation that burn hydrocarbon fuels for energy, ships create air pollution that degrades air quality, adversely affects human health and contributes to the wide-reaching effects of climate change.

## Energy Efficiency of Transportation Modes

While responsible for the release of less greenhouse gas (GHG) per tonne-kilometre of cargo transported than other forms of transportation, ships contributed 2.2% of the world's total $CO_2$ emissions in 2012.

In coming years, global shipping traffic is expected to grow in response to increased trade. Unless additional measures to limit emissions from ships are adopted, GHG emissions from shipping could increase by as much as 20% to 120% by 2050, depending on economic conditions.

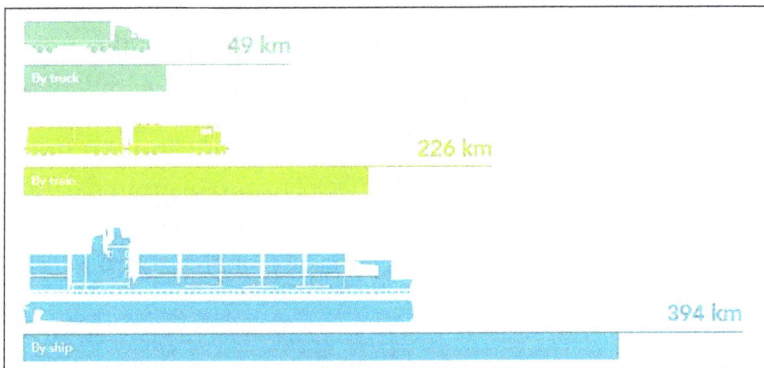

Distance 1 tonne of cargo can travel on 1 litre of fuel in Canada's Great Lakes and St. Lawrence Seaway.

## Types of Air Pollution from Marine Shipping

Commercial ships burn fuel for energy and emit several types of air pollution as by-products. Ship-source pollutants most closely linked to climate change and public health impacts include carbon dioxide ($CO_2$), nitrogen oxides ($NO_x$), sulphur oxides ($SO_x$) and particulate matter.

## Carbon Dioxide ($CO_2$)

A major GHG contributing to climate change and ocean acidification.

- $CO_2$ contributes to widespread climate change by trapping the sun's heat. In Canada, these climate changes include increased average and extreme temperatures, shifting rainfall patterns, thawing permafrost, and increases in hazardous weather.

- Climate change-induced extreme weather events such as heat waves, floods and major storms have a negative impact on human health and cause untimely deaths worldwide.

- When $CO_2$ is absorbed by seawater, the water becomes more acidic. This increase in acidity has adverse effects on marine life and ecosystems.

## Nitrogen Oxides ($NO_x$)

A collection of gases of various combinations of nitrogen and oxygen that:

- Cause lung inflammation when breathed, increasing susceptibility to harm from allergens in people with asthma. $NO_x$ may enter the bloodstream and with long-term exposure lead to eventual heart and lung failures.

- Interact with volatile organic compounds (VOCs) to create ground-level ozone, which contributes to eye, nose and throat irritations; shortness of breath; worsening of respiratory conditions; chronic obstructive pulmonary disease; asthma and allergies; cardiovascular disease and untimely death.

- Cause acidification of soil and water (acid rain).

- Decrease crop and vegetation productivity due to ground-level ozone, threatening food security.

- Flood ecosystems with excess nitrogen nutrients, leading to toxic algal blooms in coastal waters and inland lakes.

## Sulphur Oxides ($SO_x$)

A collection of gases of various combinations of sulphur and oxygen that:

- Cause lung inflammation when breathed, increasing susceptibility to allergens in people with asthma. $SO_x$ and may enter the bloodstream and with long-term exposure lead to eventual heart and lung failures.

- Cause eye irritation, increased susceptibility to respiratory tract infections, and increased hospital admissions for cardiac disease.

- Cause acidification of soil and water (acid rain).

## Particulate Matter

A collection of solid and liquid particles formed during fuel combustion that:

- Can be inhaled into people's lungs and then absorbed into the bloodstream, which has been linked to many negative heart and lung health outcomes, including cancers.

- Are a component of smog.

- Form "black carbon", the second largest contributor to climate change after $CO_q$. While airborne, black carbon absorbs solar energy and contributes to atmospheric warming, before falling to earth as precipitation that darkens snow and ice surfaces. High concentrations of black carbon on ice and snow significantly reduce solar energy reflected back into space – the albedo effect – and accelerate melting.

# Natural Sources

## Wildfire

The extreme wildfires sweeping across parts of North America, Europe and Siberia this year are not only wreaking local damage and sending choking smoke downwind. They are also affecting the climate itself in important ways that will long outlast their flames.

Wildfires emit carbon dioxide and other greenhouse gases that will continue to warm the planet well into the future. They damage forests that would otherwise remove $CO_2$ from the air. And they inject soot and other aerosols into the atmosphere, with complex effects on warming and cooling.

To be sure, the leading cause of global warming remains overwhelmingly the burning of fossil fuels. That warming lengthens the fire season, drying and heating the forests. In turn, blazes like those scorching areas across the Northern Hemisphere this summer have a feedback effect—a vicious cycle when the results of warming produce yet more warming.

### How Bad is the Climate Feedback from Fires?

Although the exact quantities are difficult to calculate, scientists estimate that wildfires emitted about 8 billion tons of $CO_2$ per year for the past 20 years. In 2017, total global $CO_2$ emissions reached 32.5 billion tons, according to the International Energy Agency.

When they calculate total global $CO_2$ output, scientists don't include all wildfire emissions as net emissions, though, because some of the $CO_2$ is offset by renewed forest growth in the burned areas. As a result, they estimate that wildfires make up 5 to 10 percent of annual global $CO_2$ emissions each year.

There have always been big wildfires, since long before humans began profoundly altering the climate by burning fossil fuels. Those historical emissions are part of the planet's natural carbon cycle. But human activities, including firefighting practices, are resulting in bigger, more intense fires, and their emissions could become a bigger contributor to global warming.

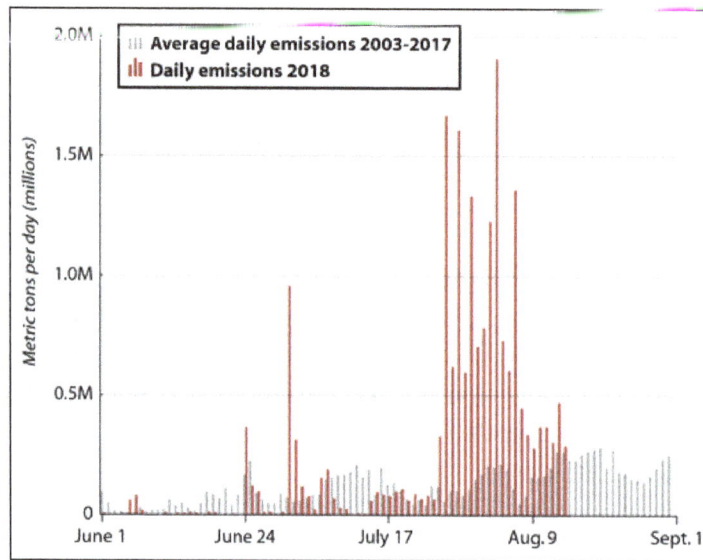

Extreme fires can release huge amounts of $CO_2$ in a very short time. Fire experts estimate that the blazes that devastated Northern California's wine country in October 2017 emitted as much $CO_2$ in one week as all of cars and trucks do over the course of a year.

Wildfires in recent years may be releasing enough $CO_2$ to endanger the state's progress toward meeting its greenhouse gas reduction targets.

While fires have been worsening in some regions, globally the total burned area and emissions from wildfires have actually decreased over the past 20 years, said Guido van der Werf, a Dutch researcher who analyzes trends for the Global Fire Emissions Database. The global decline is because burned savannas and rainforests in the tropics are being converted to agricultural lands, which are less fire-prone.

In regions of the world drying out with global warming, like the U.S. West and the Mediterranean, however, extreme fire seasons have increased in recent years.

"If we start to see a higher level of fire activity than in the past because of global warming, they become part of a climate feedback loop," van der Werf said. That means warming causes more fires, which causes more warming. In addition to their $CO_2$ emissions, wildfires can affect the climate in other important ways.

## Dead Wood and Changes to the Land

Fires don't just burn up trees and shrubs and emit smoke. They leave behind long-lasting changes on the ground, and those changes also have effects on the climate.

Over the course of several decades after a big fire, emissions from decomposing dead wood often surpass by far the direct emissions from the fire itself. But at the same time, new growth in burned areas starts to once again take $CO_2$ from the atmosphere and store it.

Fires also change the reflectivity of the land, called albedo. As burned forest areas start to regrow, lighter-colored patches of grasses and shrubs come in first, which, because they reflect more solar radiation, can have a cooling effect until the vegetation thickens and darkens again.

Scott Denning, an atmospheric scientist at Colorado State University, says site-specific studies show that the cooling effect in northern forests can last for decades. In a tropical rainforest, on the other hand, the dark canopy can regrow within a few years.

When new trees grow fast, they can start stashing away significant amounts of carbon quickly. But some recent research suggests that global warming is preventing forest regrowth after forest fires, including along the Front Range of Colorado and in the forests of the Sierra Nevada. If that emerges as a widespread trend in the coming decades, it means less forests available to take $CO_2$ out of the atmosphere. Forests are estimated to absorb up to 30 percent of human greenhouse gas emissions.

## Aerosols' Cooling and Warming Effects

Scientists can't say for certain whether the global level of fire activity in recent years is warming or cooling the atmosphere overall. Part of the reason that they don't have a definitive answer is because, along with $CO_2$, wildfires also produce many other volatile organic particles called aerosols, including substances like black carbon and gases that form ozone.

One recent study suggests that wildfires emit three times more fine particle pollution than estimated by the Environmental Protection Agency. This pollution creates health problems, and scientists are also working to better understand its impact on the climate.

Some of those aerosols can make the atmosphere more reflective. They block sunlight, which cools the atmosphere, similar to the effect attributed to emissions from volcanic eruptions. In general, the climate effect of aerosols is short-lived, lasting from a few months to a couple of years.

But black carbon, an aerosol and short-lived climate pollutant, can actually absorb heat while floating around in the air, and that heats the atmosphere. Recent research shows that the heat-trapping potency—though it is short-lived—is much higher than previously thought, roughly two-thirds that of carbon dioxide, according to Alfred Wiedensohler, with the Leibniz Institute for Tropospheric Research.

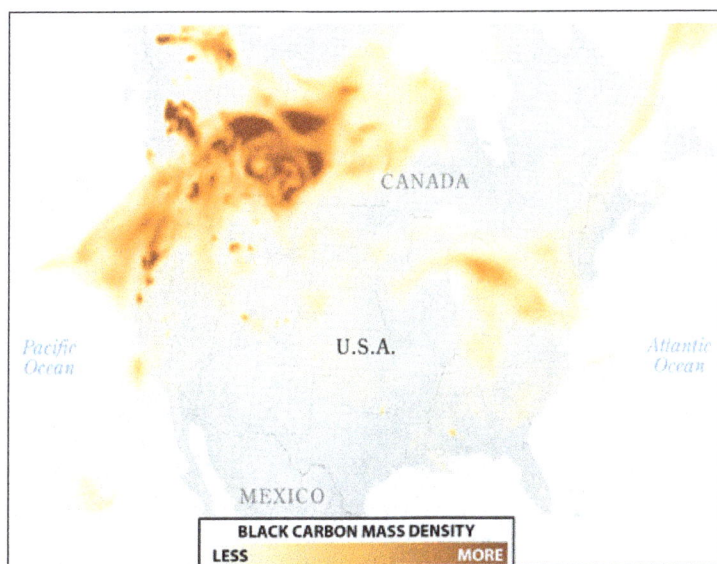

Megafires may intensify these emissions and send them higher into the atmosphere. A study published this week found that wildfires in Canada in 2017 resulted in extreme levels of aerosols over Europe, higher than those measured after the 1991 Mt. Pinatubo eruption.

An increase in megafires, driven at least partly by global warming, could change the wildfire carbon cycle, said Mark Parrington, a senior scientist with the European Centre for Medium-Range Weather Forecasts Copernicus Atmosphere Monitoring Service.

In general, if we're seeing an increase in megafires, with direct injections (of pollutants) into the upper atmosphere, the effects can linger for weeks or months, and that could have more of a climate-cooling effect.

More pieces to the wildfire-climate puzzle will fall into place after scientists evaluate data gathered by a C-130 airplane that's making daily cruises near Western U.S. wildfires to take detailed measurements of wildfire emissions. The mission is sponsored by the National Center for Atmospheric Research and the National Science Foundation.

With the explosion of wildfires in the region the past few decades, the data will help evaluate impacts to human health and the environment, including nutrient cycling, cloud formation and global warming, said University of Wyoming atmospheric scientist Shane Michael Murphy, one of the project researchers.

## Wind-blown Soot can affect the Ice Sheets

Eventually, the skies will clear once again, but all that smoke doesn't just magically disappear. The $CO_2$ will heat the atmosphere for centuries; the methane for a few decades. Some of the aerosols and other particles are heavy enough to drift earthward, and others will wash to the ground with the first good rains of autumn or winter, but not before spreading out over the Northern Hemisphere's oceans and continents.

Those tiny remnants of burned plants can also affect the climate when they land on mountain glaciers and especially on the snow and ice in the Arctic. In some years, scientists have traced soot from wildfires in Canada to Greenland, where they darken the ice and snow and speed up melting. Wildfire pollution was a significant factor in the record surface melting of the Greenland ice sheet in 2012, said climate scientist Jason Box.

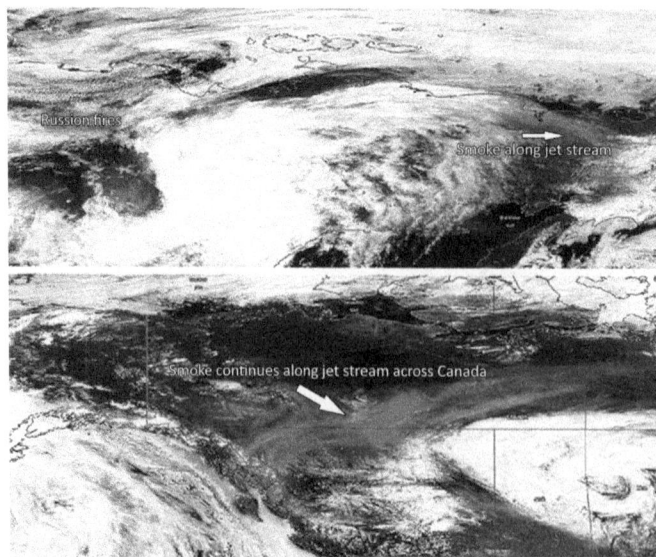

Satellite show how the jet stream spreads wildfire smoke
across Russia (top) and Canada (bottom).

The overall effect of wildfire fallout on Arctic melting is difficult to quantify, partly because of sparse sampling across the remote area, and partly because of the great annual variations in wildfire emissions. But a growing body of research suggests that wildfire soot will contribute to accelerating the Arctic meltdown in the decades ahead.

With wildfires burning farther north, emissions from wildfires in Greenland or Sweden could add significantly to the load of snow-darkening pollution in the Arctic because the sources are so close to the ice sheets. A 2016 study in Alaska estimated that risk of tundra fires will increase fourfold in the coming decades.

## Atmospheric Dust

Atmospheric dust consists of a mixture of solid and liquid particles suspened in the atmosphere varying in composition, source and size. Atmospheric dust particles can be removed out of the atmosphere by dry and wet deposition and fall back on soil, vegetation or watercourses. Atmospheric dust particles can be classified according to their diameter (measured in micrometers or µm. 1000 micrometers equivalent to 1 millimeter) ranging from 0,005 to 100 µm. Within this interval atmospheric particles are classified as:

- Primary particles – Diameter ranging from 2,5 to 30 µm;

- Secondary particles – Diameter lower than 2,5 µm.

Primary particles from combustion, soil erosion and disintegration. Pollen and spores figure in this category. Secondary particles are generated by vehicular traffic, industrial activities and thermoelectrical implants. Atmospheric dust particles with a diameter of less than 10 µm and 2,5 µm draw special attention and are defined $PM_{10}$ and $PM_{2,5}$ (PM= Particular Matter), respectively. $PM_{2,5}$ particles are a subset of $PM_{10}$ and count for 60% of its weight. $PM_{10}$ is an inhalable particle as it can travel deep into the breathing apparatus to the larynx; and it's also breathable as it can settle in the pulmonary alveoli. These dust particles raise serious health concerns as they have been linked to a number of breathing and cardiovascular diseases. Sources of dust particles can be natural (volcanic eruption, sea aerosols, spores, pollen, soil erosion) or man-made (vehicular traffic, industrial emissions and combustion processes).

## Smoke

Smoke is a collection of airborne solid and liquid particulates and gases emitted when a material undergoes combustion or pyrolysis, together with the quantity of air that is entrained or otherwise mixed into the mass. It is commonly an unwanted by-product of fires (including stoves, candles, internal combustion engines, oil lamps, and fireplaces), but may also be used for pest control (fumigation), communication (smoke signals), defensive and offensive capabilities in the military (smoke screen), cooking, or smoking (tobacco, cannabis, etc.). It is used in rituals where incense, sage, or resin is burned to produce a smell for spiritual or magical purposes. It can be a flavoring agent and preservative for various foodstuffs.

Smoke inhalation is the primary cause of death in victims of indoor fires. The smoke kills by a combination of thermal damage, poisoning and pulmonary irritation caused by carbon monoxide, hydrogen cyanide and other combustion products.

Smoke is an aerosol (or mist) of solid particles and liquid droplets that are close to the ideal range of sizes for Mie scattering of visible light.

## Chemical Composition

The composition of smoke depends on the nature of the burning fuel and the conditions of combustion. Fires with high availability of oxygen burn at a high temperature and with small amount of smoke produced; the particles are mostly composed of ash, or with large temperature differences, of condensed aerosol of water. High temperature also leads to production of nitrogen oxides. Sulfur content yields sulfur dioxide, or in case of incomplete combustion, hydrogen sulfide. Carbon and hydrogen are almost completely oxidized to carbon dioxide and water. Fires burning with lack of oxygen produce a significantly wider palette of compounds, many of them toxic. Partial oxidation of carbon produces carbon monoxide, nitrogen-containing materials can yield hydrogen cyanide, ammonia, and nitrogen oxides. Hydrogen gas can be produced instead of water. Content of halogens such as chlorine (e.g. in polyvinyl chloride or brominated flame retardants) may lead to production of e.g. hydrogen chloride, phosgene, dioxin, and chloromethane, bromomethane and other halocarbons. Hydrogen fluoride can be formed from fluorocarbons, whether fluoropolymers subjected to fire or halocarbon fire suppression agents. Phosphorus and antimony oxides and their reaction products can be formed from some fire retardant additives, increasing smoke toxicity and corrosivity. Pyrolysis of polychlorinated biphenyls (PCB), e.g. from burning older transformer oil, and to lower degree also of other chlorine-containing materials, can produce 2,3,7,8-tetrachlorodibenzodioxin, a potent carcinogen, and other polychlorinated dibenzodioxins. Pyrolysis of fluoropolymers, e.g. teflon, in presence of oxygen yields carbonyl fluoride (which hydrolyzes readily to HF and $CO_2$); other compounds may be formed as well, e.g. carbon tetrafluoride, hexafluoropropylene, and highly toxic perfluoroisobutene (PFIB).

Emission of soot in the fumes of a large diesel truck, without particle filters.

Pyrolysis of burning material, especially incomplete combustion or smoldering without adequate oxygen supply, also results in production of a large amount of hydrocarbons, both aliphatic (methane, ethane, ethylene, acetylene) and aromatic (benzene and its derivates, polycyclic aromatic hydrocarbons; e.g. benzo[a]pyrene, studied as a carcinogen, or retene), terpenes. Heterocyclic compounds may be also present. Heavier hydrocarbons may condense as tar; smoke with significant tar content is yellow to brown. Presence of such smoke, soot, and/or brown oily deposits during a fire indicates a possible hazardous situation, as the atmosphere may be saturated with combustible pyrolysis products with concentration above the upper flammability limit, and sudden inrush of air can cause flashover or backdraft.

Presence of sulfur can lead to formation of e.g. hydrogen sulfide, carbonyl sulfide, sulfur dioxide, carbon disulfide, and thiols; especially thiols tend to get adsorbed on surfaces and produce a lingering odor even long after the fire. Partial oxidation of the released hydrocarbons yields in a wide palette of other compounds: aldehydes (e.g. formaldehyde, acrolein, and furfural), ketones, alcohols (often aromatic, e.g. phenol, guaiacol, syringol, catechol, and cresols), carboxylic acids (formic acid, acetic acid, etc.).

The visible particulate matter in such smokes is most commonly composed of carbon (soot). Other particulates may be composed of drops of condensed tar, or solid particles of ash. The presence of metals in the fuel yields particles of metal oxides. Particles of inorganic salts may also be formed, e.g. ammonium sulfate, ammonium nitrate, or sodium chloride. Inorganic salts present on the surface of the soot particles may make them hydrophilic. Many organic compounds, typically the aromatic hydrocarbons, may be also adsorbed on the surface of the solid particles. Metal oxides can be present when metal-containing fuels are burned, e.g. solid rocket fuels containing aluminium. Depleted uranium projectiles after impacting the target ignite, producing particles of uranium oxides. Magnetic particles, spherules of magnetite-like ferrous ferric oxide, are present in coal smoke; their increase in deposits after 1860 marks the beginning of the Industrial Revolution. (Magnetic iron oxide nanoparticles can be also produced in the smoke from meteorites burning in the atmosphere.) Magnetic remanence, recorded in the iron oxide particles, indicates the strength of Earth's magnetic field when they were cooled beyond their Curie temperature; this can be used to distinguish magnetic particles of terrestrial and meteoric origin. Fly ash is composed mainly of silica and calcium oxide. Cenospheres are present in smoke from liquid hydrocarbon fuels. Minute metal particles produced by abrasion can be present in engine smokes. Amorphous silica particles are present in smokes from burning silicones; small proportion of silicon nitride particles can be formed in fires with insufficient oxygen. The silica particles have about 10 nm size, clumped to 70-100 nm aggregates and further agglomerated to chains. Radioactive particles may be present due to traces of uranium, thorium, or other radionuclides in the fuel; hot particles can be present in case of fires during nuclear accidents (e.g. Chernobyl disaster) or nuclear war.

Smoke particulates, like other aerosols, are categorized into three modes based on particle size:

- Nuclei mode, with geometric mean radius between 2.5–20 nm, likely forming by condensation of carbon moieties.

- Accumulation mode, ranging between 75–250 nm and formed by coagulation of nuclei mode particles.

- Coarse mode, with particles in micrometer range.

Most of the smoke material is primarily in coarse particles. Those undergo rapid dry precipitation, and the smoke damage in more distant areas outside of the room where the fire occurs is therefore primarily mediated by the smaller particles.

Aerosol of particles beyond visible size is an early indicator of materials in a preignition stage of a fire. Burning of hydrogen-rich fuel produces water; this results in smoke containing droplets of water vapor. In absence of other color sources (nitrogen oxides, particulates), such smoke is white and cloud-like.

Smoke emissions may contain characteristic trace elements. Vanadium is present in emissions from oil fired power plants and refineries; oil plants also emit some nickel. Coal combustion produces emissions containing aluminium, arsenic, chromium, cobalt, copper, iron, mercury, selenium, and uranium.

Traces of vanadium in high-temperature combustion products form droplets of molten vanadates. These attack the passivation layers on metals and cause high temperature corrosion, which is a concern especially for internal combustion engines. Molten sulfate and lead particulates also have such effect.

Some components of smoke are characteristic of the combustion source. Guaiacol and its derivatives are products of pyrolysis of lignin and are characteristic of wood smoke; other markers are syringol and derivates, and other methoxy phenols. Retene, a product of pyrolysis of conifer trees, is an indicator of forest fires. Levoglucosan is a pyrolysis product of cellulose. Hardwood vs softwood smokes differ in the ratio of guaiacols/syringols. Markers for vehicle exhaust include polycyclic aromatic hydrocarbons, hopanes, steranes, and specific nitroarenes (e.g. 1-nitropyrene). The ratio of hopanes and steranes to elemental carbon can be used to distinguish between emissions of gasoline and diesel engines.

Many compounds can be associated with particulates; whether by being adsorbed on their surfaces, or by being dissolved in liquid droplets. Hydrogen chloride is well absorbed in the soot particles.

Inert particulate matter can be disturbed and entrained into the smoke. Of particular concern are particles of asbestos.

Deposited hot particles of radioactive fallout and bioaccumulated radioisotopes can be reintroduced into the atmosphere by wildfires and forest fires; this is a concern in e.g. the Zone of alienation containing contaminants from the Chernobyl disaster.

Polymers are a significant source of smoke. Aromatic side groups, e.g. in polystyrene, enhance generation of smoke. Aromatic groups integrated in the polymer backbone produce less smoke, likely due to significant charring. Aliphatic polymers tend to generate the least smoke, and are non-self-extinguishing. However presence of additives can significantly increase smoke formation. Phosphorus-based and halogen-based flame retardants decrease production of smoke. Higher degree of cross-linking between the polymer chains has such effect too.

## Visible and Invisible Particles of Combustion

The naked eye detects particle sizes greater than 7 μm (micrometres). Visible particles emitted from a fire are referred to as smoke. Invisible particles are generally referred to as gas or fumes. This is best illustrated when toasting bread in a toaster. As the bread heats up, the products of combustion increase in size. The fumes initially produced are invisible but become visible if the toast is burnt.

Smoke from a wildfire.

An ionization chamber type smoke detector is technically a product of combustion detector, not a smoke detector. Ionization chamber type smoke detectors detect particles of combustion that are invisible to the naked eye. This explains why they may frequently false alarm from the fumes emitted from the red-hot heating elements of a toaster, before the presence of visible smoke, yet they may fail to activate in the early, low-heat smoldering stage of a fire.

Smoke rising up from the smoldering remains of a recently
extingished mountain fire.

Smoke from a typical house fire contains hundreds of different chemicals and fumes. As a result, the damage caused by the smoke can often exceed that caused by the actual heat of the fire. In addition to the physical damage caused by the smoke of a fire – which manifests itself in the form of stains – is the often even harder to eliminate problem of a smoky odor. Just as there are contractors that specialize in rebuilding/repairing homes that have been damaged by fire and smoke, fabric restoration companies specialize in restoring fabrics that have been damaged in a fire.

## Dangers

Smoke from oxygen-deprived fires contains a significant concentration of compounds that are flammable. A cloud of smoke, in contact with atmospheric oxygen, therefore has the potential of being ignited – either by another open flame in the area, or by its own temperature. This leads to effects like backdraft and flashover. Smoke inhalation is also a danger of smoke that can cause serious injury and death.

Processing fish while being exposed to smoke.

Many compounds of smoke from fires are highly toxic and/or irritating. The most dangerous is carbon monoxide leading to carbon monoxide poisoning, sometimes with the additive effects of hydrogen cyanide and phosgene. Smoke inhalation can therefore quickly lead to incapacitation and loss of consciousness. Sulfur oxides, hydrogen chloride and hydrogen fluoride in contact with moisture form sulfuric, hydrochloric and hydrofluoric acid, which are corrosive to both lungs and materials. When asleep the nose does not sense smoke nor does the brain, but the body will wake up if the lungs become enveloped in smoke and the brain will be stimulated and the person will be awoken. This does not work if the person is incapacitated or under the influence of drugs and alcohol.

Reduced visibility due to wildfire smoke in Sheremetyevo airport.

Cigarette smoke is a major modifiable risk factor for lung disease, heart disease, and many cancers. Smoke can also be a component of ambient air pollution due to the burning of coal in power plants, forest fires or other sources, although the concentration of pollutants in ambient air is typically much less than that in cigarette smoke. One day of exposure to PM2.5 at a concentration of 880 $\mu g/m^3$, such as occurs in Beijing, China, is the equivalent of smoking one or two cigarettes in terms of particulate inhalation by weight. The analysis is complicated, however, by the fact that the organic compounds present in various ambient particulates may have a higher carcinogenicity than the compounds in cigarette smoke particulates. Secondhand tobacco smoke is the combination of both sidestream and mainstream smoke emissions from a burning tobacco product. These emissions contain more than 50 carcinogenic chemicals. According to the Surgeon General's 2006 report on the subject, "Short exposures to second-hand [tobacco] smoke can cause blood platelets to become stickier, damage the lining of blood vessels, decrease coronary flow velocity reserves, and reduce heart variability, potentially increasing the risk of a heart attack". The American Cancer Society lists "heart disease, lung infections, increased asthma attacks, middle ear infections, and low birth weight" as ramifications of smoker's emission.

Red smoke carried by a parachutist of the UK Lightning
Bolts Army Parachute Display Team.

Smoke can obscure visibility, impeding occupant exiting from fire areas. In fact, the poor visibility due to the smoke that was in the Worcester Cold Storage Warehouse fire in Worcester, Massachusetts was the reason why the trapped rescue firefighters couldn't evacuate the building in time. Because of the striking similarity that each floor shared, the dense smoke caused the firefighters to become disoriented.

## Corrosion

Smoke contains a wide variety of chemicals, many of them aggressive in nature. Examples are hydrochloric acid and hydrobromic acid, produced from halogen-containing plastics and fire retardants, hydrofluoric acid released by pyrolysis of fluorocarbon fire suppression agents, sulfuric acid from burning of sulfur-containing materials, nitric acid from high-temperature fires where nitrous oxide gets formed, phosphoric acid and antimony compounds from P and Sb based fire retardants, and many others. Such corrosion is not significant for structural materials, but delicate structures, especially microelectronics, are strongly affected. Corrosion of circuit board traces, penetration of aggressive chemicals through the casings of parts, and other effects can cause an immediate or gradual deterioration of parameters or even premature (and often delayed, as the corrosion can progress over long time) failure of equipment subjected to smoke. Many smoke components are also electrically conductive; deposition of a conductive layer on the circuits can cause crosstalks and other deteriorations of the operating parameters or even cause short circuits and total failures. Electrical contacts can be affected by corrosion of surfaces, and by deposition of soot and other conductive particles or nonconductive layers on or across the contacts. Deposited particles may adversely affect the performance of optoelectronics by absorbing or scattering the light beams.

Corrosivity of smoke produced by materials is characterized by the corrosion index (CI), defined as material loss rate (angstrom/minute) per amount of material gasified products (grams) per volume of air (m³). It is measured by exposing strips of metal to flow of combustion products in a test tunnel. Polymers containing halogen and hydrogen (polyvinyl chloride, polyolefins with halogenated additives, etc.) have the highest CI as the corrosive acids are formed directly with water produced by the combustion, polymers containing halogen only (e.g. polytetrafluoroethylene) have lower CI as the formation of acid is limited to reactions with airborne humidity, and halogen-free materials (polyolefins, wood) have the lowest CI. However, some halogen-free materials can also release significant amount of corrosive products.

Smoke damage to electronic equipment can be significantly more extensive than the fire itself. Cable fires are of special concern; low smoke zero halogen materials are preferable for cable insulation.

When smoke comes into contact with the surface of any substance or structure, the chemicals contained in it are transferred to it. The corrosive properties of the chemicals cause the substance or structure to decompose at a rapid rate. Certain materials or structures absorb these chemicals, which is why clothing, unsealed surfaces, potable water, piping, wood, etc., are replaced in most cases of structural fires.

## Measurement

As early as the 15th century Leonardo da Vinci commented at length on the difficulty of assessing smoke, and distinguished between black smoke (carbonized particles) and white 'smoke' which is not a smoke at all but merely a suspension of harmless water particulates.

Smoke from heating appliances is commonly measured in one of the following ways:

- In-line capture: A smoke sample is simply sucked through a filter which is weighed before and after the test and the mass of smoke found. This is the simplest and probably the most accurate method, but can only be used where the smoke concentration is slight, as the filter can quickly become blocked.

  The ASTM smoke pump is a simple and widely used method of in-line capture where a measured volume of smoke is pulled through a filter paper and the dark spot so formed is compared with a standard.

- Filter/dilution tunnel: A smoke sample is drawn through a tube where it is diluted with air, the resulting smoke/air mixture is then pulled through a filter and weighed. This is the internationally recognized method of measuring smoke from combustion.

- Electrostatic precipitation: The smoke is passed through an array of metal tubes which contain suspended wires. A (huge) electrical potential is applied across the tubes and wires so that the smoke particles become charged and are attracted to the sides of the tubes. This method can over-read by capturing harmless condensates, or under-read due to the insulating effect of the smoke. However, it is the necessary method for assessing volumes of smoke too great to be forced through a filter, i.e., from bituminous coal.

- Ringelmann scale: A measure of smoke color. Invented by Professor Maximilian Ringelmann in Paris in 1888, it is essentially a card with squares of black, white and shades of gray which is held up and the comparative grayness of the smoke judged. Highly dependent on light conditions and the skill of the observer it allocates a grayness number from 0 (white) to 5 (black) which has only a passing relationship to the actual quantity of smoke. Nonetheless, the simplicity of the Ringelmann scale means that it has been adopted as a standard in many countries.

- Optical scattering: A light beam is passed through the smoke. A light detector is situated at an angle to the light source, typically at 90°, so that it receives only light reflected from passing particles. A measurement is made of the light received which will be higher as the concentration of smoke particles becomes higher.

- Optical obscuration: A light beam is passed through the smoke and a detector opposite measures the light. The more smoke particles are present between the two, the less light will be measured.

- Combined optical methods: There are various proprietary optical smoke measurement devices such as the 'nephelometer' or the 'aethalometer' which use several different optical methods, including more than one wavelength of light, inside a single instrument and apply an algorithm to give a good estimate of smoke. It has been claimed that these devices can differentiate types of smoke and so their probable source can be inferred, though this is disputed.

- Inference from carbon monoxide: Smoke is incompletely burned fuel, carbon monoxide is incompletely burned carbon, therefore it has long been assumed that measurement of CO in flue gas (a cheap, simple and very accurate procedure) will provide a good indication of the levels of smoke. Indeed, several jurisdictions use CO measurement as the basis of smoke control. However it is far from clear how accurate the correspondence is.

## Health Benefits

Throughout recorded history, humans have used the smoke of medicinal plants to cure illness. A sculpture from Persepolis shows Darius the Great (522–486 BC), the king of Persia, with two censers in front of him for burning Peganum harmala and/or sandalwood Santalum album, which was believed to protect the king from evil and disease. More than 300 plant species in 5 continents are used in smoke form for different diseases. As a method of drug administration, smoking is important as it is a simple, inexpensive, but very effective method of extracting particles containing active agents. More importantly, generating smoke reduces the particle size to a microscopic scale thereby increasing the absorption of its active chemical principles.

## Volcanic Eruptions

A volcano is an open fissure on the surface of the earth. Active volcanoes are those from which lava, volcanic ashes, rocks, dust and gas compounds escape on a regular basis (10.000 years are considered regular with volcanoes, so you can feel safe if you have one around) due to the phenomenon of volcanic eruptions.

In the world there are several active volcanoes which cause air pollution, danger to life forms and massive destruction of the land and the environment. Indonesia is the country with most active volcanoes in the world with 76 of them and total of 147 volcanoes.

## Causes of Volcanic Eruptions

There are many causes that can lead to a volcanic emissions, many of them are still unknown by humans, which volcano eruptions very hard to predict. However, volcanologists have made some researchers to determine a few catalysts of them:

- Movement of tectonic plates: Whether it is because one is pushed under another one or two tectonic plates are moved away from each other, this creates a massive movement on the layers of planet earth (changing the structures of magma, sediments and seawater) and cause a volcano to erupt.

- Decreasing temperatures: The volume of magma changes when it crystalizes, so it can push away liquid magma and create a volcanic eruption.

- Decrease in external pressure: This fact provokes an increase in the internal pressure of the volcano and causes and eruption if it is not capable of holding back the lava.

- Buoyancy of the magma: If the density of the magma between the zone of its generation and the surface is less than that of the surrounding and overlying rocks, the magma reaches the surface and erupts.

- Pressure from the exsolved gases: Andesitic and rhyolitic magma compositions contain dissolved volatiles (gases) such as water, sulfur dioxide and carbon dioxide. This gas bubbles are held by magma, but just like a carbonated drink, the bubbles of gas rise to the surface of the magma chamber creating a volcanic eruption.

- Injection of a new batch of magma into an already filled magma chamber: This phenomenon causes some magma to move up and spill or even erupt at the surface.

## Effects of Volcanic Eruptions

Volcanoes have huge impact on our society and environment when they erupt, here there are some of the many positive and negative effects of volcanic eruptions:

- Negative effects: Volcanic eruptions, which sometimes generate earthquakes, can destroy landscapes, natural resources, wildlife and human lives and their properties. This phenomenon can also discharge ashes very high into the atmosphere, having negative consequences on the ozone layer. Moreover, ash and mud can mix with rain and melting snow and create situations like lahars (also called mudflows) or acid rain. In other words, volcanic eruptions can destroy civilizations, like what happened to Pompeii.

- Positive Effects: Sometimes eruptions can leave an extraordinary beautiful and natural scenery, attacking tourists to the area. However, one of the most useful positive effects is that they often leave potential for geothermal energy, making life more easy for those around the area. Finally, some volcanic eruptions provide valuable nutrients for the soil, which are later used as fertile soils for agriculture.

## Types of Volcanic Eruptions

## Air Pollution of Volcanic Eruptions

Volcanic Eruptions release massive quantities of solid pollutants and gases, forming enormous clouds that can affect areas miles away from the volcanic eruption. Therefore, volcanoes are an international form of air pollution, but not just for us, as a lot of this greenhouse gases and aerosols go directly into the atmosphere.

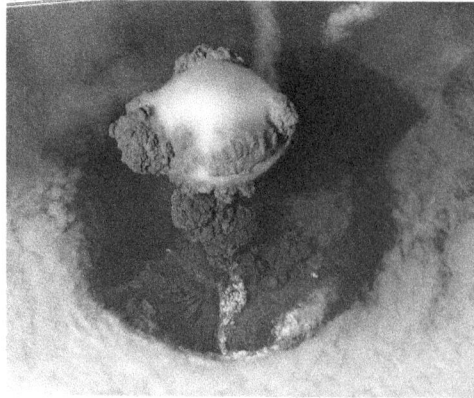

Every volcanic eruption is different on impact, and therefore different on the quantity and a variety of pollutants emitted. On average the outgassed composition release is 79% water vapor ($H_2O$), 11,6% carbon dioxide ($CO_2$), 6,5% sulphur dioxide ($SO_2$) and 2,9% of other pollutants.

However, the range of pollutants released on a volcanic eruption include: Carbon dioxide ($CO_2$), sulphur dioxide ($SO_2$), hydrogen sulfide ($H_2S$), hydrogen ($H_2$), hydrogen fluoride (HF), hydrogen chloride (HCl), bromide oxide (BrO) and carbon monoxide (CO). Highly exposure to these gases has detrimental impact on living organisms both terrestrial and marine.

On the other side, particulates are another source of air pollution produced by volcanic eruptions. Mainly ashes and including all types of sizes, the ones that help forming toxic clouds are usually $PM_{10}$, $PM_{2.5}$, $PM_{0.3}$ and thinner.

These pollution clouds can travel major distances, like crossing oceans, dangerously affecting people and environments who didn't even notice the eruption. If you are facing a polluted cloud caused by a volcanic eruption, you may want to have a look at the following content on what to do and how to prevent these events.

## References

- Seinfeld, john; spyros pandis (1998). Atmospheric chemistry and physics: from air pollution to climate change (2nd ed.). Hoboken, new jersey: john wiley & sons. P. 97. Isbn 978-0-471-17816-3

- Toxic-air-pollutants, air-pollution, outdoor, healthy-air, our-initiatives: lung.org, Retrieved 21 April, 2019

- Nox-pollution, nox-problem: icopal-noxite.co.uk, Retrieved 24 March, 2019

- Seinfeld, John H.; Pandis, Spyros N. (2016). Atmospheric Chemistry and Physics: From Air Pollution to Climate Change (3rd ed.). Wiley. ISBN 978-1-119-22116-6

- Kaufman, y. J.; fraser, robert s. (1997). "the effect of smoke particles on clouds and climate forcing". Science. 277 (5332): 1636–1639. Doi:10.1126/science.277.5332.1636

- Bond, T. C. (2013). "Bounding the role of black carbon in the climate system: A scientific assessment". Journal of Geophysical Research: Atmospheres. 118 (11): 5380–5552. Bibcode:2013JGRD..118.5380B. doi:10.1002/jgrd.50171

- Volatile-organic-compounds-impact-indoor-air-quality, indoor-air-quality-iaq: epa.gov, Retrieved 12 April, 2019

- Chemistry-emission-control-radioactive-pollution-and-indoor-air-quality/indoor-air-pollutants-and-the-impact-on-human-health, books: intechopen.com, Retrieved 23 June, 2019

# 3
# Effects of Air Pollution

Air pollution adversely affects the health in humans, especially harming the lungs and brain. Some of its other effects include arctic haze, global warming, ozone depletion and acid rain. The topics elaborated in this chapter will help in gaining a better perspective about these effects of air pollution.

## Health Effects of Air Pollution

Air pollution is one of the world's largest killers, responsible for 6.4 million deaths per year (1 in 9 deaths), of which 600,000 are children. This is more than the number of deaths from AIDS, Malaria and tuberculosis combined. The World Health Organization estimates that 2 billion children live in areas where outdoor air pollution exceeds international limits and 300 million children live in areas where outdoor air pollution exceeds 6 times international limits. Children, the elderly, and people with heart or lung disease, diabetes, minority and low - income communities are particularly vulnerable to adverse health outcomes from exposure to air pollution, including cardiovascular disease, asthma and other respiratory diseases, and cancer. Recent evidence suggests that air pollution is also linked to higher risk of diabetes, autism, and lower IQ.

What we typically think of as "air pollution" is actually a mixture of small particles (such as: black carbon gases like nitrogen oxides, ozone, and sulfur dioxide.

### Particulate Matter ($PM_{10}$, $PM_{2.5}$)

Particulate matter (PM) is made up of small airborne particles like dust, soot, and drops of liquids. The majority of PM in urban areas is formed directly from burning of fossil fuels by power plants, automobiles, non-road equipment, and industrial facilities. Other sources are dust and diesel emissions and secondary particle formation from gases and vapors.

Coarse particulate matter ($PM_{10}$, particles < 10 microns in diameter) is known to cause nasal and upper respiratory tract health problems. Fine particles ($PM_{2.5}$, particles < 2.5 microns in diameter; Ultra Fine Particles) penetrate deeper into the lungs and cause heart attacks, strokes, asthma, and bronchitis, as well as premature death from heart ailments, lung disease, and cancer. Studies show that higher $PM_{2.5}$ exposure can impair brain development in children.

## Black Carbon (BC)

Black carbon is one of the components of particulate matter and comes from burning fuel (especially diesel, wood, coal, and others). Most air pollution regulations focus on $PM_{2.5}$, but exposure to black carbon is a serious health threat as well. Populations with higher exposures to black carbon over a long period are at a higher risk for heart attacks and stroke. In addition, black carbon is associated with hypertension, asthma, chronic obstructive pulmonary disease (COPD), bronchitis, and a variety of types of cancer.

## Nitrogen Oxides (NO and $NO_2$)

Nitrogen oxide (NO) and Nitrogen dioxide ($NO_2$) are produced primarily by the transportation sector. NO is rapidly converted to $NO_2$ in sunlight. $NO_x$ (a combination of NO and $NO_2$) is formed in high concentrations around roadways, and can result in development and exacerbations of asthma, bronchitis, as well as lead to a higher risk of heart disease.

## Ozone ($O_3$)

Ozone high up in the atmosphere can protect us from ultraviolet radiation. But ozone at ground level (where it is part of what is commonly called smog) is a well-established respiratory irritant. Ozone is formed in the atmosphere through reactions of volatile organic compounds and nitrogen oxides, both of which are formed as a result of combustion of fossil fuels. Short-term exposure to ozone can cause chest pain, coughing, throat irritation, while long term exposure can lead to decreased lung function and cause chronic obstructive pulmonary disease (COPD). In addition, ozone exposure can aggravate existing lung diseases.

## Sulfur dioxide ($SO_2$)

$SO_2$ is emitted into the air by the burning of fossil fuels that contain sulfur. Coal, metal extraction and smelting, ship engines, and heavy equipment diesel equipment burn fuels that contain sulfur. Sulfur dioxide causes eye irritation, worsens asthma, increases susceptibility to respiratory infections, and impacts the cardiovascular system. When $SO_2$ combines with water, it forms sulfuric acid; this is the main component of acid rain, a known contributor to deforestation.

## Air Pollution and the Brain

A great deal of scientific research demonstrates that air pollution exposure can lead to harmful health effects related to the lungs and heart. More recent studies show that air pollution can have harmful effects on the brain.

Studies over the past two decades suggest that air pollution exposures are linked to harmful effects on the brain.

The children had poorer brain health than the other children, including the following:

- Breakdown of the brain's protective layer;
- Changes in the brain resembling the early stages of Alzheimer's disease; and
- Poorer performance on standardized psychological tests.

# Can Air Pollutants enter or affect the Brain?

Studies that exposed animals to ultrafine particulate matter (UFPM, particles less than 0.1 micrometers in diameter) showed the following:

- Inhaled particles can travel through a nerve running from the nasal cavity into the brain, bypassing the lungs.

- UFPM carried in the bloodstream can pass through the barrier that normally protects the brain.

- Chemical markers of brain inflammation can increase.

Thus, air pollutants can affect the brain directly and indirectly, and are associated with potentially harmful effects. These findings strengthen the case for inhaled pollutant-related brain impacts.

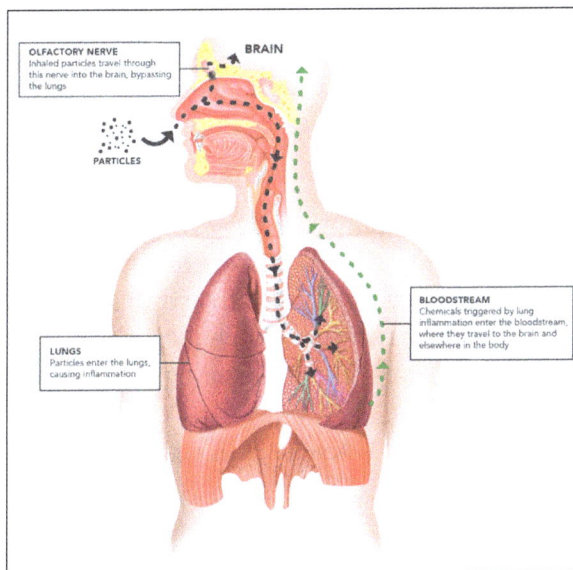

Potential pathways by which inhaled particles affect the brain.

# Can Air Pollution affect Cognition and Brain Disorders?

Animals exposed to particulate matter or diesel particulates demonstrated:

- Poorer performance in mazes, and other learning and memory problems.

- Behaviors resembling human anxiety, depression, and impulsiveness.

In humans, emerging evidence suggests links between air pollution and harmful brain effects in the elderly. Two recent studies revealed associations between air pollution and dementia:

- A study in Ontario, Canada found that the risk of dementia increased the closer people lived to major roadways.

- A study of U.S. women showed higher risk of cognitive decline and dementia for those exposed to levels of fine particulate matter (PM$_{2.5}$, particles less than 2.5 micrometers in diameter) above the national standard.

## Effects of Air Pollution on Lungs

Air pollution exposure can trigger new cases of asthma, exacerbate (worsen) a previously-existing respiratory illness, and provoke development or progression of chronic illnesses including lung cancer, chronic obstructive pulmonary disease, and emphysema. Air pollutants also negatively and significantly harm lung development, creating an additional risk factor for developing lung diseases later in life.

## Asthma

Asthma, a chronic disease of the lungs characterized by inflammation and narrowing of the airways, causes a sensation of tightness in the chest, shortness of breath, wheezing, and coughing. If untreated, asthma episodes can be nearfatal or even fatal. Asthma is not currently curable, and damage that is done to lung tissue during asthma attacks may lead to permanent damage. Nearly 1.8 million emergency room visits were attributed to asthma in 2005. There are many triggers to asthma attacks, including dust, smoke, pollen, and volatile organic compounds. Common outdoor pollutant triggers include ozone, carbon monoxide, sulfur dioxide and nitrogen oxides.

## Asthma-ozone Connection

Ozone is formed when volatile organic compounds react with nitrogen oxides in the presence of sunlight. Ozone irritates the lungs at concentrations which are fairly common in urban settings, particularly in summer months. Increases in ozone are linked to asthma and other lung diseases. For those with severe asthma, symptoms increase even when ambient ozone levels fall under the thresholds set by the EPA. Elevated ozone levels also aggravate pre-existing heart problems, like angina.

## Chronic Obstructive Pulmonary Disease, Chronic Bronchitis and Emphysema

Chronic Obstructive Pulmonary Disease (COPD) is another condition characterized by narrowing of the airways, but these changes are permanent rather than reversible. COPD is caused by exposure to pollutants that produce inflammation, an immunological response. In larger airways, the inflammatory response is referred to as chronic bronchitis. In the tiny air cells at the end of the lung's smallest passageways, it leads to destruction of tissue, or emphysema. Although current and ex-smokers account for most patients with COPD, exposure to air pollutants plays an important role in the development of COPD and the origin and development of acute exacerbations.

## Lung Cancer

Lung cancer, the leading U.S. cancer killer in both men and women, is often (and accurately) associated with smoking tobacco. While that's true, there are multiple other risk factors for developing lung cancer, including air pollution. Particulate matter and ozone in particular may affect mortality due to lung cancer.

## Children are Especially Vulnerable

Children are particularly susceptible to the effects of air pollution. They breathe through their mouths, bypassing the filtering effects of the nasal passages and allowing pollutants to travel deeper into the lungs. They have a large lung surface area relative to their weight and inhale relatively more air, compared to adults. They also spend more time out of doors, particularly in the afternoons and during the summer months when ozone and other pollutant levels are at their highest. And, children may ignore early symptoms of air pollution effects, such as an asthma exacerbation, leading to attacks of increased severity. Combine those factors with the adverse impact of some pollutants on lung development and the immaturity of children's enzyme and immune systems that detoxify pollutants, and you have a series of factors that contribute to children's increased sensitivity to air pollutants.

## Health Effects of Indoor Air Pollution

The effects of indoor air pollutants range from short-term effects – eye and throat irritation – to long-term effects – respiratory disease and cancer. Exposure to high levels of some pollutants, such as carbon monoxide, can even result in immediate death. Also, some indoor pollutants can magnify the effects of other indoor pollutants.

## Common Symptoms of Indoor Air Pollution

Symptoms of poor indoor air quality are very broad and depend on the contaminant. They can easily be mistaken for symptoms of other illnesses such as allergies, stress, colds and influenza. The most common symptoms are:

- Coughing
- Sneezing
- Watery eyes
- Fatigue
- Dizziness
- Headaches
- Upper respiratory congestion

If you notice relief from your symptoms soon after leaving a particular room or building, your symptoms may be caused by indoor air contaminants.

## Respiratory Health Effects

- Rhinitis, nasal congestion (inflammation of the nose, runny nose)
- Epistaxis (nose bleeds)
- Dyspnea (difficulty of breathing or painful breathing)
- Pharyngitis (sore throat), cough
- Wheezing, worsening asthma
- Severe lung disease

## More Severe Health Effects

- Conjunctival (eye) irritation

- Rashes

- Fever, chills

- Tachycardia (rapid heartbeat, sometimes leading to shortness of breath)

- Headache or dizziness

- Lethargy, fatigue, malaise

- Nausea, vomiting, anorexia

- Myalgia (muscle pain)

- Hearing loss

# Air Stagnation

Air stagnation is a phenomenon which occurs when an air mass remains over an area for an extended period. Stagnation events strongly correlates with poor air quality. Due to light winds and lack of precipitation, pollutants cannot be cleared from the air, either gaseous (such as ozone) or particulate (such as soot or dust). Subsidence produced directly under the subtropical ridge can lead to a buildup of particulates in urban areas under the ridge, leading to widespread haze. If the low level relative humidity rises towards 100 percent overnight, fog can form. In the United States, the National Weather Service issues an *Air Stagnation Advisory* when these conditions are likely to occur.

## Air Stagnation Index

Atmospheric pollution manifests itself in many ways, ranging from reduced visibility to dangerous respiratory problems and discomfort. Atmospheric pollution can be gaseous (e.g. ozone, sulfur dioxide, nitrogen oxides) and/or particulate (e.g. soot, dust). The degree of pollution is dependent on a number of factors: source, transport from source, and temporal build up through stagnation. The stagnation index is intended as an indication of the latter only.

In this context, stagnation is considered to consist of light winds so that horizontal dispersion is at a minimum, a stable lower atmosphere that effectively prevents vertical escape, and no precipitation to wash any pollution away. These conditions are most frequently met when there is a persistent or slow moving high pressure system.

One of the most prevalent air pollutants present in the lower levels of the atmosphere during the late spring and summer seasons is ground level ozone. This ozone is produced when pollutants from cars and factories are "cooked" by a hot summer sun. Consequently, concentrations of ground level ozone are typically highest during periods of high temperature and become a health problem particularly when air is stagnant.

During the summer of 2002, conditions from Missouri east to New Jersey and down through South Carolina were conducive to air stagnation. Compared to the summer of 2000, the percentage of days with air stagnation conditions nearly doubled. The maps below show conditions in July 2000 and July 2002.

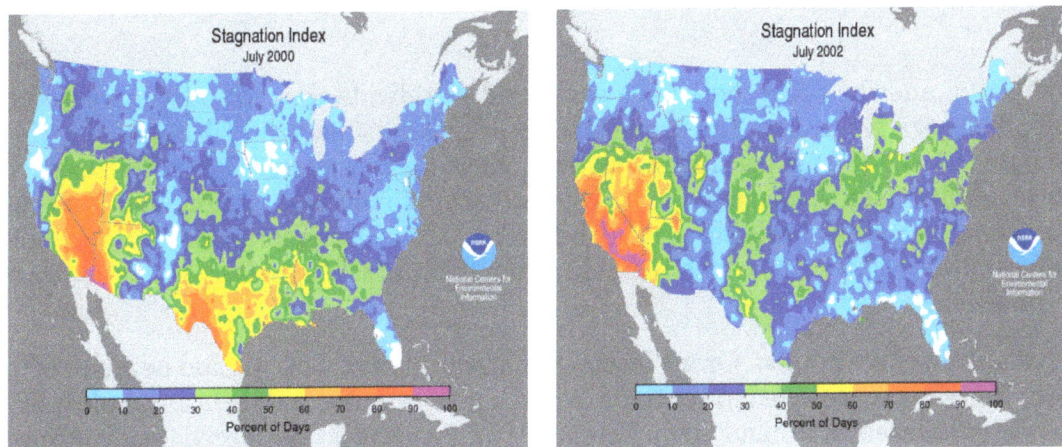

The yellow areas indicate a moderate 8-hour average peak concentration whereas the orange and red areas indicate locations where the concentration of ozone was deemed unhealthy.

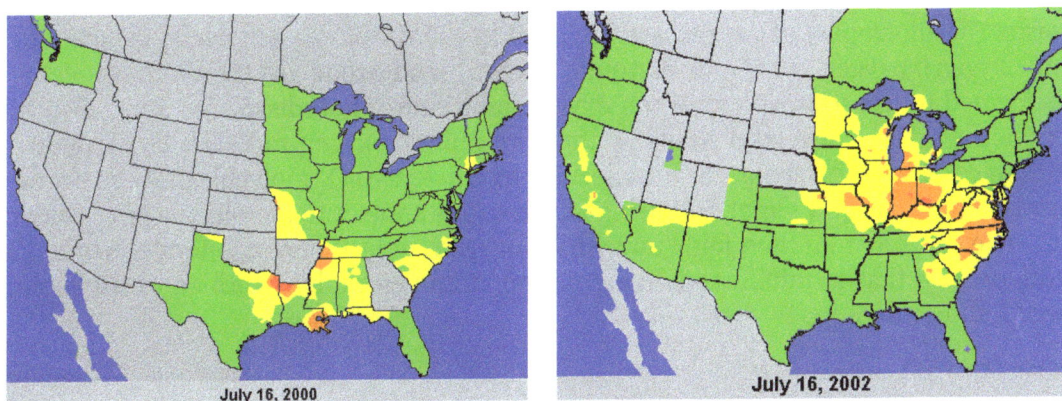

The Air Stagnation Index maps provided here are based on the work of Wang and Angell, but with some slight modifications. To make assessment of stagnation totally objective, they defined a stagnation day as one with sea level geostrophic wind less than 8m/sec, 500mb wind less than 13m/sec, and no precipitation. If there is a temperature inversion below 850mb the 8m/sec is relaxed by 10% (to 8.8m/sec). We calculated a stagnation day using the same method but performed the calculations over a finer grid (0.25 x 0.25 degree instead of 2.5 x 2.5 degree). The Wang and Angell stagnation index is the number of 4-day stagnation periods.

# Ozone Depletion

The main cause of ozone depletion and the ozone hole is manufactured chemicals, especially manufactured halocarbon refrigerants, solvents, propellants and foam-blowing agents (chlorofluorocarbons (CFCs), HCFCs, halons), referred to as ozone-depleting substances (ODS). These compounds are transported into the stratosphere by turbulent mixing after being emitted from the surface, mixing much faster than the molecules can settle. Once in the stratosphere, they release halogen atoms through photodissociation, which catalyze the breakdown of ozone ($O_3$) into oxygen ($O_2$). Both types of ozone depletion were observed to increase as emissions of halocarbons increased.

Ozone depletion and the ozone hole have generated worldwide concern over increased cancer risks and other negative effects. The ozone layer prevents most harmful UVB wavelengths of ultraviolet light (UV light) from passing through the Earth's atmosphere. These wavelengths cause skin cancer, sunburn and cataracts, which were projected to increase dramatically as a result of thinning ozone, as well as harming plants and animals. These concerns led to the adoption of the Montreal Protocol in 1987, which bans the production of CFCs, halons and other ozone-depleting chemicals.

The ban came into effect in 1989. Ozone levels stabilized by the mid-1990s and began to recover in the 2000s. Recovery is projected to continue over the next century, and the ozone hole is expected to reach pre-1980 levels by around 2075. The Montreal Protocol is considered the most successful international environmental agreement to date.

## Ozone Cycle

Three forms (or allotropes) of oxygen are involved in the ozone-oxygen cycle: oxygen atoms (O or atomic oxygen), oxygen gas ($O_2$ or diatomic oxygen), and ozone gas ($O_3$ or triatomic oxygen). Ozone is formed in the stratosphere when oxygen molecules photodissociate after absorbing ultraviolet photons. This converts a single $O_2$ into two atomic oxygen radicals. The atomic oxygen radicals then combine with separate $O_2$ molecules to create two $O_3$ molecules. These ozone molecules absorb ultraviolet (UV) light, following which ozone splits into a molecule of $O_2$ and an oxygen atom. The oxygen atom then joins up with an oxygen molecule to regenerate ozone. This is a continuing process that terminates when an oxygen atom recombines with an ozone molecule to make two $O_2$ molecules.

$$O + O_3 \rightarrow 2O_2$$

The total amount of ozone in the stratosphere is determined by a balance between photochemical production and recombination.

Ozone can be destroyed by a number of free radical catalysts; the most important are the hydroxyl radical (OH·), nitric oxide radical (NO·), chlorine radical (Cl·) and bromine radical (Br·). The dot is a notation to indicate that each species has an unpaired electron and is thus extremely reactive. All of these have both natural and man-made sources; at the present time, most of the OH· and NO· in the stratosphere is naturally occurring, but human activity has drastically increased the levels of chlorine and bromine. These elements are found in stable organic compounds, especially chlorofluorocarbons, which can travel to the stratosphere without being destroyed in the troposphere due to their low reactivity. Once in the stratosphere, the Cl and Br atoms are released from the parent compounds by the action of ultraviolet light, e.g.:

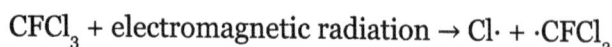

$$CFCl_3 + \text{electromagnetic radiation} \rightarrow Cl\cdot + \cdot CFCl_2$$

Ozone is a highly reactive molecule that easily reduces to the more stable oxygen form with the assistance of a catalyst. Cl and Br atoms destroy ozone molecules through a variety of catalytic cycles. In the simplest example of such a cycle, a chlorine atom reacts with an ozone molecule ($O_3$), taking an oxygen atom to form chlorine monoxide (ClO) and leaving an oxygen molecule ($O_2$). The ClO can react with a second molecule of ozone, releasing the chlorine atom and yielding two molecules of oxygen. The chemical shorthand for these gas-phase reactions is:

$$Cl\cdot + O_3 \rightarrow ClO + O_2$$

A chlorine atom removes an oxygen atom from an ozone molecule to make a ClO molecule.

$$ClO + O_3 \rightarrow Cl\cdot + 2\, O_2$$

This ClO can also remove an oxygen atom from another ozone molecule; the chlorine is free to repeat this two-step cycle.

The overall effect is a decrease in the amount of ozone, though the rate of these processes can be decreased by the effects of null cycles. More complicated mechanisms have also been discovered that lead to ozone destruction in the lower stratosphere.

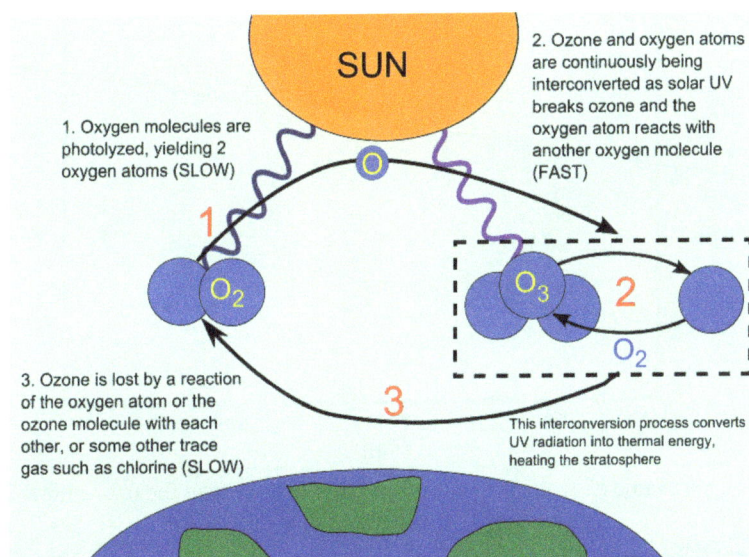

The ozone cycle.

A single chlorine atom would continuously destroy ozone (thus a catalyst) for up to two years (the time scale for transport back down to the troposphere) were it not for reactions that remove them from this cycle by forming reservoir species such as hydrogen chloride (HCl) and chlorine nitrate ($ClONO_2$). Bromine is even more efficient than chlorine at destroying ozone on a per atom basis, but there is much less bromine in the atmosphere at present. Both chlorine and bromine contribute significantly to overall ozone depletion. Laboratory studies have also shown that fluorine and iodine atoms participate in analogous catalytic cycles. However, fluorine atoms react rapidly with water and methane to form strongly bound HF in the Earth's stratosphere, while organic molecules containing iodine react so rapidly in the lower atmosphere that they do not reach the stratosphere in significant quantities.

A single chlorine atom is able to react with an average of 100,000 ozone molecules before it is removed from the catalytic cycle. This fact plus the amount of chlorine released into the atmosphere yearly by chlorofluorocarbons (CFCs) and hydrochlorofluorocarbons (HCFCs) demonstrates the danger of CFCs and HCFCs to the environment.

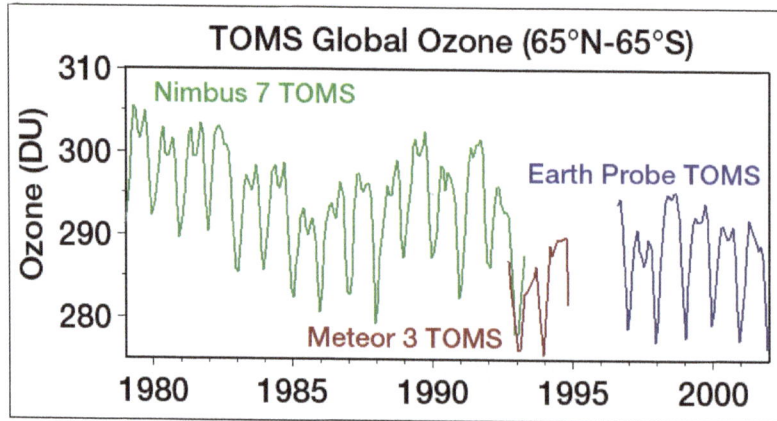
Global monthly average total ozone amount.

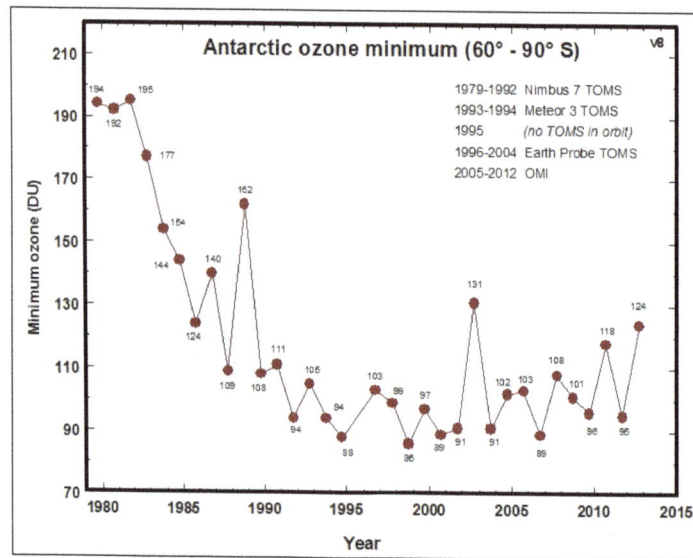
Lowest value of ozone measured by TOMS each year in the ozone hole.

## Observations on Ozone Layer Depletion

The ozone hole is usually measured by reduction in the total *column ozone* above a point on the Earth's surface. This is normally expressed in Dobson units; abbreviated as "DU". The most prominent decrease in ozone has been in the lower stratosphere. Marked decreases in column ozone in the Antarctic spring and early summer compared to the early 1970s and before have been observed using instruments such as the Total Ozone Mapping Spectrometer (TOMS).

Reductions of up to 70 percent in the ozone column observed in the austral (southern hemispheric) spring over Antarctica and first reported in 1985 (Farman et al.) are continuing. Antarctic total column ozone in September and October have continued to be 40–50 percent lower than pre-ozone-hole values since the 1990s. A gradual trend toward "healing" was reported in 2016. In 2017, NASA announced that the ozone hole was the weakest since 1988 because of warm stratospheric conditions. It is expected to recover around 2070.

The amount lost is more variable year-to-year in the Arctic than in the Antarctic. The greatest Arctic declines are in the winter and spring, reaching up to 30 percent when the stratosphere is coldest.

Reactions that take place on polar stratospheric clouds (PSCs) play an important role in enhancing ozone depletion. PSCs form more readily in the extreme cold of the Arctic and Antarctic stratosphere. This is why ozone holes first formed, and are deeper, over Antarctica. Early models failed to take PSCs into account and predicted a gradual global depletion, which is why the sudden Antarctic ozone hole was such a surprise to many scientists.

It is more accurate to speak of ozone depletion in middle latitudes rather than holes. Total column ozone declined below pre-1980 values between 1980 and 1996 for mid-latitudes. In the northern mid-latitudes, it then increased from the minimum value by about two percent from 1996 to 2009 as regulations took effect and the amount of chlorine in the stratosphere decreased. In the Southern Hemisphere's mid-latitudes, total ozone remained constant over that time period. There are no significant trends in the tropics, largely because halogen-containing compounds have not had time to break down and release chlorine and bromine atoms at tropical latitudes.

Large volcanic eruptions have been shown to have substantial albeit uneven ozone-depleting effects, as observed with the 1991 eruption of Mt. Pinotubo in the Philippines.

Ozone depletion also explains much of the observed reduction in stratospheric and upper tropospheric temperatures. The source of the warmth of the stratosphere is the absorption of UV radiation by ozone, hence reduced ozone leads to cooling. Some stratospheric cooling is also predicted from increases in greenhouse gases such as $CO_2$ and CFCs themselves; however, the ozone-induced cooling appears to be dominant.

## Compounds in the Atmosphere

### CFCs and related Compounds in the Atmosphere

Chlorofluorocarbons (CFCs) and other halogenated ozone depleting substances (ODS) are mainly responsible for man-made chemical ozone depletion. The total amount of effective halogens (chlorine and bromine) in the stratosphere can be calculated and are known as the equivalent effective stratospheric chlorine (EESC).

CFCs were invented by Thomas Midgley, Jr. in the 1930s. They were used in air conditioning and cooling units, as aerosol spray propellants prior to the 1970s, and in the cleaning processes of delicate electronic equipment. They also occur as by-products of some chemical processes. No significant natural sources have ever been identified for these compounds—their presence in the atmosphere is due almost entirely to human manufacture. As mentioned above, when such ozone-depleting chemicals reach the stratosphere, they are dissociated by ultraviolet light to release chlorine atoms. The chlorine atoms act as a catalyst, and each can break down tens of thousands of ozone molecules before being removed from the stratosphere. Given the longevity of CFC molecules, recovery times are measured in decades. It is calculated that a CFC molecule takes an average of about five to seven years to go from the ground level up to the upper atmosphere, and it can stay there for about a century, destroying up to one hundred thousand ozone molecules during that time.

1,1,1-Trichloro-2,2,2-trifluoroethane, also known as CFC-113a, is one of four man-made chemicals newly discovered in the atmosphere by a team at the University of East Anglia. CFC-113a is the only known CFC whose abundance in the atmosphere is still growing. Its source remains a mystery, but illegal manufacturing is suspected by some. CFC-113a seems to have been accumulating unabated since 1960. Between 2010 and 2012, emissions of the gas jumped by 45 percent.

## Computer Modeling

Scientists have attributed ozone depletion to the increase of man-made (anthropogenic) halogen compounds from CFCs by combining observational data with computer models. These complex chemistry transport models (e.g. SLIMCAT, CLaMS—Chemical Lagrangian Model of the Stratosphere) work by combining measurements of chemicals and meteorological fields with chemical reaction rate constants. They identify key chemical reactions and transport processes that bring CFC photolysis products into contact with ozone.

## Ozone Hole and its Causes

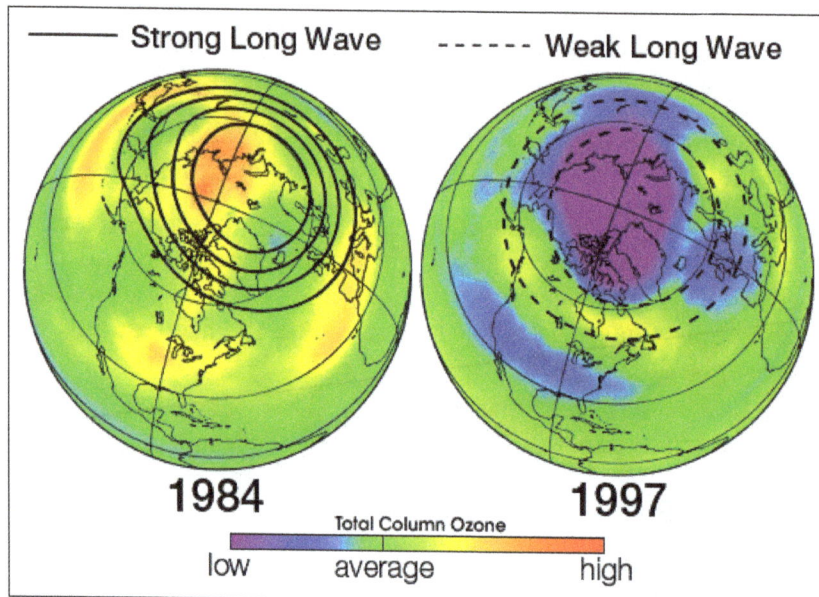

Ozone hole in North America during 1984 (abnormally warm reducing ozone depletion) and 1997 (abnormally cold resulting in increased seasonal depletion).

The Antarctic ozone hole is an area of the Antarctic stratosphere in which the recent ozone levels have dropped to as low as 33 percent of their pre-1975 values. The ozone hole occurs during the Antarctic spring, from September to early December, as strong westerly winds start to circulate around the continent and create an atmospheric container. Within this polar vortex, over 50 percent of the lower stratospheric ozone is destroyed during the Antarctic spring.

As explained above, the primary cause of ozone depletion is the presence of chlorine-containing source gases (primarily CFCs and related halocarbons). In the presence of UV light, these gases dissociate, releasing chlorine atoms, which then go on to catalyze ozone destruction. The Cl-catalyzed ozone depletion can take place in the gas phase, but it is dramatically enhanced in the presence of polar stratospheric clouds (PSCs).

These polar stratospheric clouds form during winter, in the extreme cold. Polar winters are dark, consisting of three months without solar radiation (sunlight). The lack of sunlight contributes to a decrease in temperature and the polar vortex traps and chills air. Temperatures hover around or below −80 °C. These low temperatures form cloud particles. There are three types of PSC clouds—nitric acid trihydrate clouds, slowly cooling water-ice clouds, and rapid cooling water-ice (nacreous) clouds—provide surfaces for chemical reactions whose products will, in the spring lead to ozone destruction.

The photochemical processes involved are complex but well understood. The key observation is that, ordinarily, most of the chlorine in the stratosphere resides in "reservoir" compounds, primarily chlorine nitrate ($ClONO_2$) as well as stable end products such as HCl. The formation of end products essentially remove Cl from the ozone depletion process. The former sequester Cl, which can be later made available via absorption of light at shorter wavelengths than 400 nm. During the Antarctic winter and spring, however, reactions on the surface of the polar stratospheric cloud particles convert these "reservoir" compounds into reactive free radicals (Cl and ClO). The process by which the clouds remove $NO_2$ from the stratosphere by converting it to nitric acid in the PSC particles, which then are lost by sedimentation is called denitrification. This prevents newly formed ClO from being converted back into $ClONO_2$.

The role of sunlight in ozone depletion is the reason why the Antarctic ozone depletion is greatest during spring. During winter, even though PSCs are at their most abundant, there is no light over the pole to drive chemical reactions. During the spring, however, the sun comes out, providing energy to drive photochemical reactions and melt the polar stratospheric clouds, releasing considerable ClO, which drives the hole mechanism. Further warming temperatures near the end of spring break up the vortex around mid-December. As warm, ozone and $NO_2$- rich air flows in from lower latitudes, the PSCs are destroyed, the enhanced ozone depletion process shuts down, and the ozone hole closes.

Most of the ozone that is destroyed is in the lower stratosphere, in contrast to the much smaller ozone depletion through homogeneous gas phase reactions, which occurs primarily in the upper stratosphere.

## Interest in Ozone Layer Depletion

Public misconceptions and misunderstandings of complex issues like the ozone depletion are common. The limited scientific knowledge of the public led to a confusion with global warming or the perception of global warming as a subset of the "ozone hole". In the beginning, classical green NGOs refrained from using CFC depletion for campaigning, as they assumed the topic was too complicated. They became active much later, e.g. in Greenpeace's support for a CFC-free fridge produced by the former East German company VEB dkk Scharfenstein.

The metaphors used in the CFC discussion (ozone shield, ozone hole) are not "exact" in the scientific sense. The "ozone hole" is more of a *depression*, less "a hole in the windshield". The ozone does not disappear through the layer, nor is there a uniform "thinning" of the ozone layer. However they resonated better with non-scientists and their concerns. The ozone hole was seen as a "hot issue" and imminent risk as lay people feared severe personal consequences such as skin cancer, cataracts, damage to plants, and reduction of plankton populations in the ocean's photic zone. Not only on the policy level, ozone regulation compared to climate change fared much better in public opinion. Americans voluntarily switched away from aerosol sprays before legislation was enforced, while climate change failed to achieve comparable concern and public action. The sudden recognition in 1985 that there was a substantial "hole" was widely reported in the press. The especially rapid ozone depletion in Antarctica had previously been dismissed as a measurement error. Scientific consensus was established after regulation.

While the Antarctic ozone hole has a relatively small effect on global ozone, the hole has generated a great deal of public interest because:

- Many have worried that ozone holes might start appearing over other areas of the globe, though to date the only other large-scale depletion is a smaller ozone "dimple" observed during the Arctic spring around the North Pole. Ozone at middle latitudes has declined, but by a much smaller extent (a decrease of about 4–5 percent).

• If stratospheric conditions become more severe (cooler temperatures, more clouds, more active chlorine), global ozone may decrease at a greater pace. Standard global warming theory predicts that the stratosphere will cool.

• When the Antarctic ozone hole breaks up each year, the ozone-depleted air drifts out into nearby regions. Decreases in the ozone level of up to 10 percent have been reported in New Zealand in the month following the breakup of the Antarctic ozone hole, with ultraviolet-B radiation intensities increasing by more than 15 percent since the 1970s.

## Consequences of Ozone Layer Depletion

Since the ozone layer absorbs UVB ultraviolet light from the sun, ozone layer depletion increases surface UVB levels (all else equal), which could lead to damage, including increase in skin cancer. This was the reason for the Montreal Protocol. Although decreases in stratospheric ozone are well-tied to CFCs and to increases in surface UVB, there is no direct observational evidence linking ozone depletion to higher incidence of skin cancer and eye damage in human beings. This is partly because UVA, which has also been implicated in some forms of skin cancer, is not absorbed by ozone, and because it is nearly impossible to control statistics for lifestyle changes over time.

## Increased UV

Ozone, while a minority constituent in Earth's atmosphere, is responsible for most of the absorption of UVB radiation. The amount of UVB radiation that penetrates through the ozone layer decreases exponentially with the slant-path thickness and density of the layer. When stratospheric ozone levels decrease, higher levels of UVB reach the Earth's surface. UV-driven phenolic formation in tree rings has dated the start of ozone depletion in northern latitudes to the late 1700s.

In October 2008, the Ecuadorian Space Agency published a report called HIPERION. The study used ground instruments in Ecuador and the last 28 years' data from 12 satellites of several countries, and found that the UV radiation reaching equatorial latitudes was far greater than expected, with the UV Index climbing as high as 24 in Quito; the WHO considers 11 as an extreme index and a great risk to health. The report concluded that depleted ozone levels around the mid-latitudes of the planet are already endangering large populations in these areas. Later, the CONIDA, the Peruvian Space Agency, published its own study, which yielded almost the same findings as the Ecuadorian study.

## Biological Effects

The main public concern regarding the ozone hole has been the effects of increased surface UV radiation on human health. So far, ozone depletion in most locations has been typically a few percent and, as noted above, no direct evidence of health damage is available in most latitudes. If the high levels of depletion seen in the ozone hole were to be common across the globe, the effects could be substantially more dramatic. As the ozone hole over Antarctica has in some instances grown so large as to affect parts of Australia, New Zealand, Chile, Argentina, and

South Africa, environmentalists have been concerned that the increase in surface UV could be significant.

Ozone depletion would magnify all of the effects of UV on human health, both positive (including production of vitamin D) and negative (including sunburn, skin cancer, and cataracts). In addition, increased surface UV leads to increased tropospheric ozone, which is a health risk to humans.

## Basal and Squamous Cell Carcinomas

The most common forms of skin cancer in humans, basal and squamous cell carcinomas, have been strongly linked to UVB exposure. The mechanism by which UVB induces these cancers is well understood—absorption of UVB radiation causes the pyrimidine bases in the DNA molecule to form dimers, resulting in transcription errors when the DNA replicates. These cancers are relatively mild and rarely fatal, although the treatment of squamous cell carcinoma sometimes requires extensive reconstructive surgery. By combining epidemiological data with results of animal studies, scientists have estimated that every one percent decrease in long-term stratospheric ozone would increase the incidence of these cancers by two percent.

## Malignant Melanoma

Another form of skin cancer, malignant melanoma, is much less common but far more dangerous, being lethal in about 15–20 percent of the cases diagnosed. The relationship between malignant melanoma and ultraviolet exposure is not yet fully understood, but it appears that both UVB and UVA are involved. Because of this uncertainty, it is difficult to estimate the effect of ozone depletion on melanoma incidence. One study showed that a 10 percent increase in UVB radiation was associated with a 19 percent increase in melanomas for men and 16 percent for women. A study of people in Punta Arenas, at the southern tip of Chile, showed a 56 percent increase in melanoma and a 46 percent increase in nonmelanoma skin cancer over a period of seven years, along with decreased ozone and increased UVB levels.

## Cortical Cataracts

Epidemiological studies suggest an association between ocular cortical cataracts and UVB exposure, using crude approximations of exposure and various cataract assessment techniques. A detailed assessment of ocular exposure to UVB was carried out in a study on Chesapeake Bay Watermen, where increases in average annual ocular exposure were associated with increasing risk of cortical opacity. In this highly exposed group of predominantly white males, the evidence linking cortical opacities to sunlight exposure was the strongest to date. Based on these results, ozone depletion is predicted to cause hundreds of thousands of additional cataracts by 2050.

## Increased Tropospheric Ozone

Increased surface UV leads to increased tropospheric ozone. Ground-level ozone is generally recognized to be a health risk, as ozone is toxic due to its strong oxidant properties. The risks are particularly high for young children, the elderly, and those with asthma or other respiratory difficulties. At this time, ozone at ground level is produced mainly by the action of UV radiation on combustion gases from vehicle exhausts.

## Increased Production of Vitamin D

Vitamin D is produced in the skin by ultraviolet light. Thus, higher UVB exposure raises human vitamin D in those deficient in it. Recent research (primarily since the Montreal Protocol) shows that many humans have less than optimal vitamin D levels. In particular, in the U.S. population, the lowest quarter of vitamin D (<17.8 ng/ml) were found using information from the National Health and Nutrition Examination Survey to be associated with an increase in all-cause mortality in the general population. While blood level of vitamin D in excess of 100 ng/ml appear to raise blood calcium excessively and to be associated with higher mortality, the body has mechanisms that prevent sunlight from producing vitamin D in excess of the body's requirements.

## Effects on Animals

A November 2010 report by scientists at the Institute of Zoology in London found that whales off the coast of California have shown a sharp rise in sun damage, and these scientists "fear that the thinning ozone layer is to blame". The study photographed and took skin biopsies from over 150 whales in the Gulf of California and found "widespread evidence of epidermal damage commonly associated with acute and severe sunburn", having cells that form when the DNA is damaged by UV radiation. The findings suggest "rising UV levels as a result of ozone depletion are to blame for the observed skin damage, in the same way that human skin cancer rates have been on the increase in recent decades."

## Effects on Crops

An increase of UV radiation would be expected to affect crops. A number of economically important species of plants, such as rice, depend on cyanobacteria residing on their roots for the retention of nitrogen. Cyanobacteria are sensitive to UV radiation and would be affected by its increase. "Despite mechanisms to reduce or repair the effects of increased ultraviolet radiation, plants have a limited ability to adapt to increased levels of UVB, therefore plant growth can be directly affected by UVB radiation."

## Prospects of Ozone Depletion

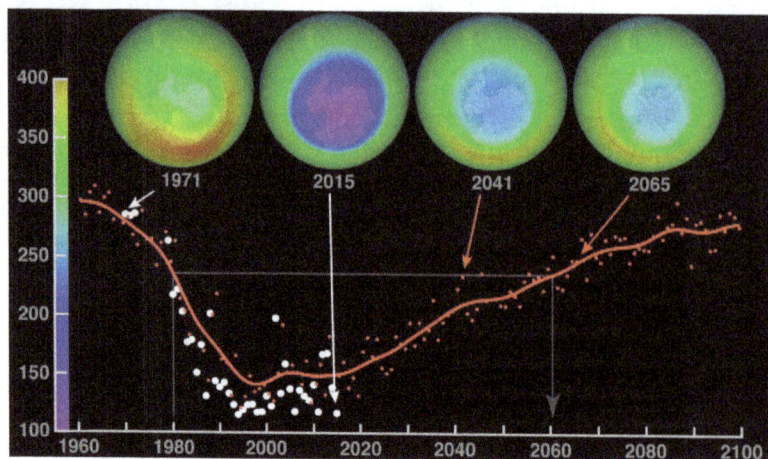

Ozone levels stabilized in the 1990s following the Montreal Protocol, and have started to recover. They are projected to reach pre-1980 levels before 2075.

Since the adoption and strengthening of the Montreal Protocol has led to reductions in the emissions of CFCs, atmospheric concentrations of the most-significant compounds have been declining. These substances are being gradually removed from the atmosphere; since peaking in 1994, the Effective Equivalent Chlorine (EECl) level in the atmosphere had dropped about 10 percent by 2008. The decrease in ozone-depleting chemicals has also been significantly affected by a decrease in bromine-containing chemicals. The data suggest that substantial natural sources exist for atmospheric methyl bromide ($CH_3Br$). The phase-out of CFCs means that nitrous oxide ($N_2O$), which is not covered by the Montreal Protocol, has become the most highly emitted ozone-depleting substance and is expected to remain so throughout the 21st century.

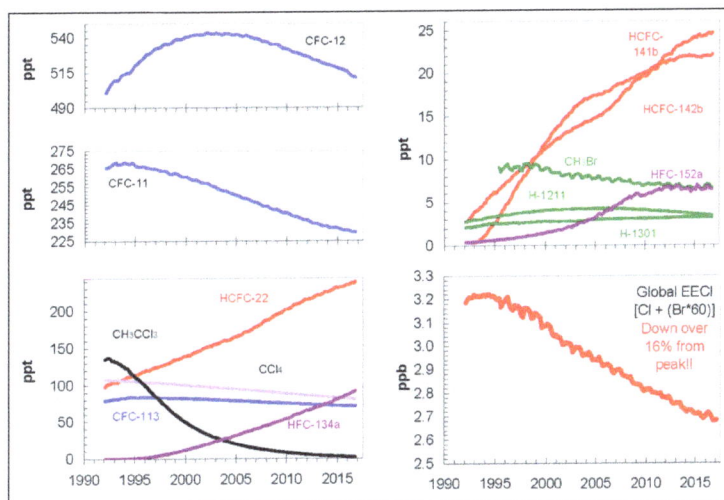

Ozone-depleting gas trends.

A 2005 IPCC review of ozone observations and model calculations concluded that the global amount of ozone has now approximately stabilized. Although considerable variability is expected from year to year, including in polar regions where depletion is largest, the ozone layer is expected to begin to recover in coming decades due to declining ozone-depleting substance concentrations, assuming full compliance with the Montreal Protocol.

The Antarctic ozone hole is expected to continue for decades. Ozone concentrations in the lower stratosphere over Antarctica will increase by 5–10 percent by 2020 and return to pre-1980 levels by about 2060–2075. This is 10–25 years later than predicted in earlier assessments, because of revised estimates of atmospheric concentrations of ozone-depleting substances, including a larger predicted future usage in developing countries. Another factor that may prolong ozone depletion is the drawdown of nitrogen oxides from above the stratosphere due to changing wind patterns. A gradual trend toward "healing" was reported in 2016.

## Ozone Depletion and Global Warming

Among others, Robert Watson had a role in the science assessment and in the regulation efforts of ozone depletion and global warming. Prior to the 1980s, the EU, NASA, NAS, UNEP, WMO and the British government had dissenting scientific reports and Watson played a role in the process of unified assessments. Based on the experience with the ozone case, the IPCC started to work on a unified reporting and science assessment to reach a consensus to provide the IPCC Summary for Policymakers.

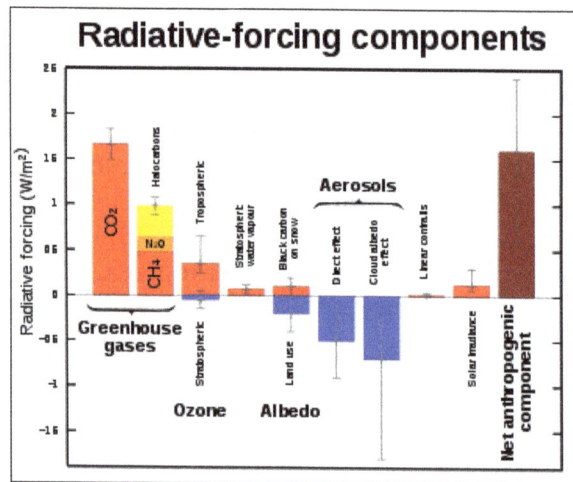

Radiative forcing from various greenhouse gases and other sources.

There are various areas of linkage between ozone depletion and global warming science:

- The same $CO_2$ radiative forcing that produces global warming is expected to cool the stratosphere. This cooling, in turn, is expected to produce a relative *increase* in ozone ($O_3$) depletion in polar area and the frequency of ozone holes.

- Conversely, ozone depletion represents a radiative forcing of the climate system. There are two opposing effects: Reduced ozone causes the stratosphere to absorb less solar radiation, thus cooling the stratosphere while warming the troposphere; the resulting colder stratosphere emits less long-wave radiation downward, thus cooling the troposphere. Overall, the cooling dominates; the IPCC concludes "observed stratospheric $O_3$ losses over the past two decades have caused a negative forcing of the surface-troposphere system" of about $-0.15 \pm 0.10$ watts per square meter ($W/m^2$).

- One of the strongest predictions of the greenhouse effect is that the stratosphere will cool. Although this cooling has been observed, it is not trivial to separate the effects of changes in the concentration of greenhouse gases and ozone depletion since both will lead to cooling. However, this can be done by numerical stratospheric modeling. Results from the National Oceanic and Atmospheric Administration's Geophysical Fluid Dynamics Laboratory show that above 20 km (12 mi), the greenhouse gases dominate the cooling.

- Ozone depleting chemicals are also often greenhouse gases. The increases in concentrations of these chemicals have produced $0.34 \pm 0.03$ $W/m^2$ of radiative forcing, corresponding to about 14 percent of the total radiative forcing from increases in the concentrations of well-mixed greenhouse gases.

- The long term modeling of the process, its measurement, study, design of theories and testing take decades to document, gain wide acceptance, and ultimately become the dominant paradigm. Several theories about the destruction of ozone were hypothesized in the 1980s, published in the late 1990s, and are currently being investigated. Dr Drew Schindell, and Dr Paul Newman, NASA Goddard, proposed a theory in the late 1990s, using computational modeling methods to model ozone destruction, that accounted for 78 percent of the ozone destroyed. Further refinement of that model accounted for 89 percent of the ozone destroyed, but pushed back the estimated recovery of the ozone hole from 75 years to 150 years. An important part of that model is the lack of stratospheric flight due to depletion of fossil fuels.

## Misconceptions

### CFC Weight

Since CFC molecules are heavier than air (nitrogen or oxygen), it is commonly believed that the CFC molecules cannot reach the stratosphere in significant amount. However, atmospheric gases are not sorted by weight; the forces of wind can fully mix the gases in the atmosphere. Lighter CFCs are evenly distributed throughout the turbosphere and reach the upper atmosphere, although some of the heavier CFCs are not evenly distributed.

### Percentage of Man-made Chlorine

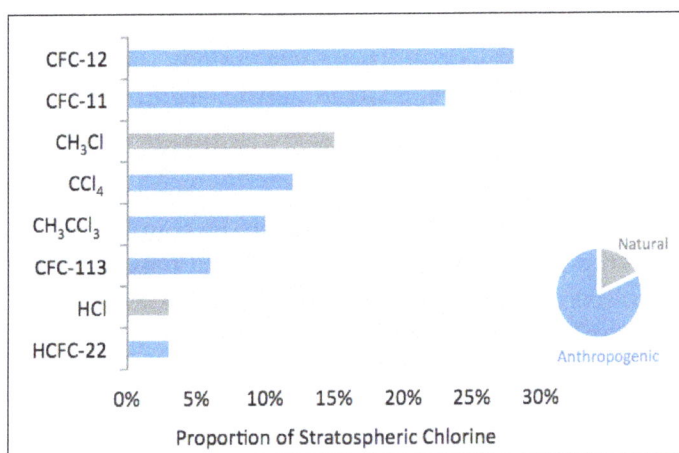

Sources of stratospheric chlorine.

Another misconception is that "it is generally accepted that natural sources of tropospheric chlorine are four to five times larger than man-made ones." While this statement is strictly true, *tropospheric* chlorine is irrelevant; it is *stratospheric* chlorine that affects ozone depletion. Chlorine from ocean spray is soluble and thus is washed by rainfall before it reaches the stratosphere. CFCs, in contrast, are insoluble and long-lived, allowing them to reach the stratosphere. In the lower atmosphere, there is much more chlorine from CFCs and related haloalkanes than there is in HCl from salt spray, and in the stratosphere halocarbons are dominant. Only methyl chloride, which is one of these halocarbons, has a mainly natural source, and it is responsible for about 20 percent of the chlorine in the stratosphere; the remaining 80 percent comes from manmade sources.

Very violent volcanic eruptions can inject HCl into the stratosphere, but researchers have shown that the contribution is not significant compared to that from CFCs. A similar erroneous assertion is that soluble halogen compounds from the volcanic plume of Mount Erebus on Ross Island, Antarctica are a major contributor to the Antarctic ozone hole.

Nevertheless, a 2015 study showed that the role of Mount Erebus volcano in the Antarctic ozone depletion was probably underestimated. Based on the NCEP/NCAR reanalysis data over the last 35 years and by using the NOAA HYSPLIT trajectory model, researchers showed that Erebus volcano gas emissions (including hydrogen chloride (HCl)) can reach the Antarctic stratosphere via high latitude cyclones and then the polar vortex. Depending on Erebus volcano activity, the additional annual HCl mass entering the stratosphere from Erebus varies from 1.0 to 14.3 kt.

## First Observation

G.M.B. Dobson mentioned that when springtime ozone levels in the Antarctic over Halley Bay were first measured in 1956, he was surprised to find that they were ~320 DU, or about 150 DU below spring Arctic levels of ~450 DU. These were at that time the only known Antarctic ozone values available. What Dobson describes is essentially the *baseline* from which the ozone hole is measured: actual ozone hole values are in the 150–100 DU range.

The discrepancy between the Arctic and Antarctic noted by Dobson was primarily a matter of timing: during the Arctic spring ozone levels rose smoothly, peaking in April, whereas in the Antarctic they stayed approximately constant during early spring, rising abruptly in November when the polar vortex broke down.

The behavior seen in the Antarctic ozone hole is completely different. Instead of staying constant, early springtime ozone levels suddenly drop from their already low winter values, by as much as 50 percent, and normal values are not reached again until December.

## Location of Hole

Some people thought that the ozone hole should be above the sources of CFCs. However, CFCs are well mixed globally in the troposphere and stratosphere. The reason for occurrence of the ozone hole above Antarctica is not because there are more CFCs concentrated but because the low temperatures help form polar stratospheric clouds. In fact, there are findings of significant and localized "ozone holes" above other parts of the earth.

## World Ozone Day

In 1994, the United Nations General Assembly voted to designate September 16 as the International Day for the Preservation of the Ozone Layer, or "World Ozone Day", to commemorate the signing of the Montreal Protocol on that date in 1987.

# Global Warming

Global warming can be defined as an increase in the average temperature of the Earth due to air pollutants, which collect sunlight and radiation and produce the greenhouse effect. This pollution layer avoids the reflection of sun rays by Earth's surface towards space, which raise the temperature in our planet among a lot more consequences.

It is a real problem since statistics and evidence are there. According to NASA's data:

- Carbon monoxide levels in the air are the highest in 650.000 years, concretely up to 408 ppm (parts per million).

- 17 out of the 18 warmest years in history (which have been recorded) have taken place after 2001. Global temperature has increased 1 °C (1,8 °F) since 1880.

- Arctic ice minimum levels have decreased 13,2% each decade. In 2012, Arctic summer sea ice shrank to the lowest extent on record.

- Satellite data show that Earth's polar ice sheets are losing mass at speed of 413 gigatonnes per year.

- Sea level is currently increasing 3, 2 millimeters per year.

## Causes of Global Warming

The main cause of global warming, according to most climate scientists, is the greenhouse effect. Greenhouse gases absorb heat and re-emit it in all directions warming up the lower atmosphere and the Earth's surface.

Burning of oil and fossil fuels, which mainly come from sources such as vehicles combustion engines and carbon industries, emit huge amounts of carbon molecules to the atmosphere, where they react with oxygen to create carbon dioxide ($CO_2$). More sources of global warming contributing to create that pollution layer "in charge of microwaving the Earth" are soot and aerosols, among many others.

## Greenhouse Gases

These gases can be classified in two groups depending if they react to changes (physically or chemically) ("feedbacks") or not ("forcing"). Greenhouse gases are:

- Water Vapor ($H_2O$): The most abundant greenhouse gas. It performs as feedback: the warmest the Earth is, more water vapor will be found in the atmosphere (clouds and precipitation).

- Nitrous oxide ($N_2O$): Human-made activities such as soil cultivation, use of fertilisers, fossil fuel combustion, nitric acid production or biomass burning.

- Methane ($CH_4$): An active, but limited greenhouse gas emitted by human and biological sources like agriculture or ruminant digestion associated to livestock.

- Carbon dioxide ($CO_2$): Although produced naturally by respiration or volcano eruptions, carbon dioxide emissions have become an environmental problem due to anthropogenic sources. Since Industrial Revolution, activities such as industrial operations, deforestation or burning of fossil fuels, among others have massively increased the emission of this greenhouse gas.

- Chlorofluorocarbons (CFCs): Synthetic compounds produced by industrial activities and controlled by some governments due to its strong effects related to the ozone layer depletion.

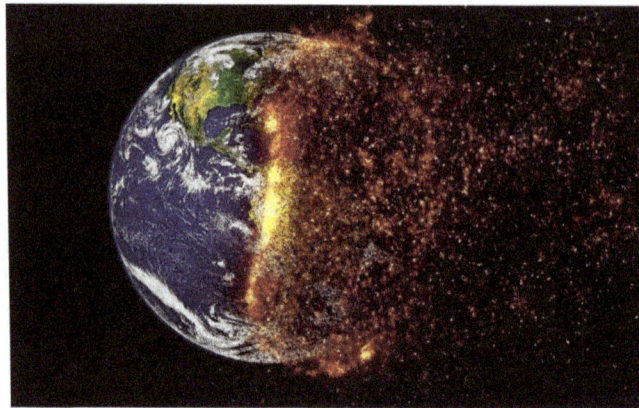

## Global Warming Environmental Effects: Climate Change

Mainly, global warming implies the worst of the air pollution environmental effects: Climate change. By warming up the surface of our planet, we get an impact almost worldwide.

Some of the current evidence of climate change are:

- Ice declining and sea level rising:
    - Ice sheets slip: The warmer the planet is, faster the ice will thaw.
    - Glacial shrink, especially in places such as the Alps, Himalayas, Andes, Rockies, Alaska and Africa.
    - Less snow covering the top of mountains.
    - Sea level rise at current speed of 3,2 millimeters per year, which in 50 years mean coastline will have risen over 15 centimeters.
    - Downturning the thickness and extension of Arctic sea ice.
- Extreme weather events:
    - Extreme events such as hurricane Katrina, which hit New Orleans and other american cities of Louisiana and Florida.
- Ecosystem changes:
    - Warmer global temperature: As already mentioned, since the 19th century the average temperature of the planet has increased by 1 °C (1,8 °F).
    - Warming oceans: oceans have collected this increased heat within the top 700 meters.
    - Ocean acidification: The acidity of surface ocean waters has increased by about 30% due to humans activities emitting carbon dioxide into the atmosphere.

However, this does not end here. The effects will continue to grow and environmental impact will rise over the years:

- Temperatures will continue increasing.

- Extending frost-free and growing seasons for crops.

- Precipitation patterns will be affected.

- Changes in natural habitats that will cost the extinction of a huge amount of plants and animals.

- More droughts and heat waves.

- More intense and stronger hurricanes.

- Sea level will rise from 30 to 120 centimeters by the end of 21th century.

- Arctic won't have ice anymore.

## Health Effects of Global Warming

Despite climate change has direct health effects in humans, most of the problems or issues it causes are due to environmental alterations. Heat waves, natural disasters, breathing poor air quality and spreading diseases are just some examples.

Some groups such as kids or elderly are more vulnerable to illness or death due to these global warming consequences. However, it affects differently depending on the region and the capacity of each country to adapt to changes.

Global warming will also affect crops, livestock, fisheries and others by reducing yields, seasonal and weather changes, the need of using more pests, etc. Furthermore, drinking water will become harder to find, less availability and of less quality.

## How to Stop and Prevent Global Warming?

Changing your daily routines is not easy, but if you expect a future we need to make the effort and change some habits. It is not only a government's duty, it is in our hands.

Some possible solutions to reduce global warming are:

- Investing in renewable energies for our homes, businesses, means of transportation, etc.

- Not use fossil fuel electricity anymore.

- Change our food production habits in order to prevent deforestation and forest degradation, as these emissions represent the 30% of the world's heat trapping emissions.

- Improving in nuclear power, so it has fewer pollution emissions.

- Improve and apply new low-carbon and zero-carbon technologies.

- Reduce water waste.

# Acid Rain

Acid rain is a rain or any other form of precipitation that is unusually acidic, meaning that it has elevated levels of hydrogen ions (low pH). It can have harmful effects on plants, aquatic animals and infrastructure. Acid rain is caused by emissions of sulfur dioxide and nitrogen oxide, which react with the water molecules in the atmosphere to produce acids. Some governments have made efforts since the 1970s to reduce the release of sulfur dioxide and nitrogen oxide into the atmosphere with positive results. Nitrogen oxides can also be produced naturally by lightning strikes, and sulfur dioxide is produced by volcanic eruptions. Acid rain has been shown to have adverse impacts on forests, freshwaters and soils, killing insect and aquatic life-forms, causing paint to peel, corrosion of steel structures such as bridges, and weathering of stone buildings and statues as well as having impacts on human health.

"Acid rain" is a popular term referring to the deposition of a mixture from wet (rain, snow, sleet, fog, cloudwater, and dew) and dry (acidifying particles and gases) acidic components. Distilled water, once carbon dioxide is removed, has a neutral pH of 7. Liquids with a pH less than 7 are acidic, and those with a pH greater than 7 are alkaline. "Clean" or unpolluted rain has an acidic pH, but usually no lower than 5.7, because carbon dioxide and water in the air react together to form carbonic acid, a weak acid according to the following reaction:

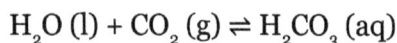

$$H_2O \ (l) + CO_2 \ (g) \rightleftharpoons H_2CO_3 \ (aq)$$

Carbonic acid then can ionize in water forming low concentrations of carbonate and hydronium ions:

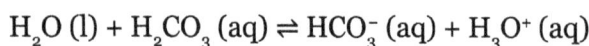

$$H_2O \ (l) + H_2CO_3 \ (aq) \rightleftharpoons HCO_3^- \ (aq) + H_3O^+ \ (aq)$$

Unpolluted rain can also contain other chemicals which affect its pH (acidity level). A common example is nitric acid produced by electric discharge in the atmosphere such as lightning. Acid deposition as an environmental issue would include additional acids other than $H_2CO_3$.

## Emissions of Chemicals Leading to Acidification

The most important gas which leads to acidification is sulfur dioxide. Emissions of nitrogen oxides which are oxidized to form nitric acid are of increasing importance due to stricter controls on emissions of sulfur compounds. 70 Tg(S) per year in the form of $SO_2$ comes from fossil fuel combustion and industry, 2.8 Tg(S) from wildfires and 7–8 Tg(S) per year from volcanoes.

## Natural Phenomena

The principal natural phenomena that contribute acid-producing gases to the atmosphere are emissions from volcanoes. Thus, for example, fumaroles from the Laguna Caliente crater of Poás Volcano create extremely high amounts of acid rain and fog, with acidity as high as a pH of 2, clearing an area of any vegetation and frequently causing irritation to the eyes and lungs of inhabitants in nearby settlements. Acid-producing gasses are also created by biological processes that occur on the land, in wetlands, and in the oceans. The major biological source of sulfur compounds is dimethyl sulfide.

Nitric acid in rainwater is an important source of fixed nitrogen for plant life, and is also produced by electrical activity in the atmosphere such as lightning.

Acidic deposits have been detected in glacial ice thousands of years old in remote parts of the globe.

Soils of coniferous forests are naturally very acidic due to the shedding of needles, and the results of this phenomenon should not be confused with acid rain.

## Human Activity

The coal-fired Gavin Power Plant in Cheshire.

The principal cause of acid rain is sulfur and nitrogen compounds from human sources, such as electricity generation, factories, and motor vehicles. Electrical power generation using coal is among the greatest contributors to gaseous pollution responsible for acidic rain. The gases can be carried hundreds of kilometers in the atmosphere before they are converted to acids and deposited. In the past, factories had short funnels to let out smoke but this caused many problems locally; thus, factories now have taller smoke funnels. However, dispersal from these taller stacks causes pollutants to be carried farther, causing widespread ecological damage.

## Chemical Processes

Combustion of fuels produces sulfur dioxide and nitric oxides. They are converted into sulfuric acid and nitric acid.

## Gas Phase Chemistry

In the gas phase sulfur dioxide is oxidized by reaction with the hydroxyl radical via an intermolecular reaction:

$$SO_2 + OH\cdot \rightarrow HOSO_2\cdot$$

which is followed by:

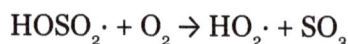

$$HOSO_2\cdot + O_2 \rightarrow HO_2\cdot + SO_3$$

In the presence of water, sulfur trioxide ($SO_3$) is converted rapidly to sulfuric acid:

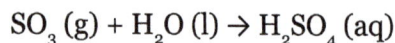

$$SO_3 (g) + H_2O (l) \rightarrow H_2SO_4 (aq)$$

Nitrogen dioxide reacts with OH to form nitric acid:

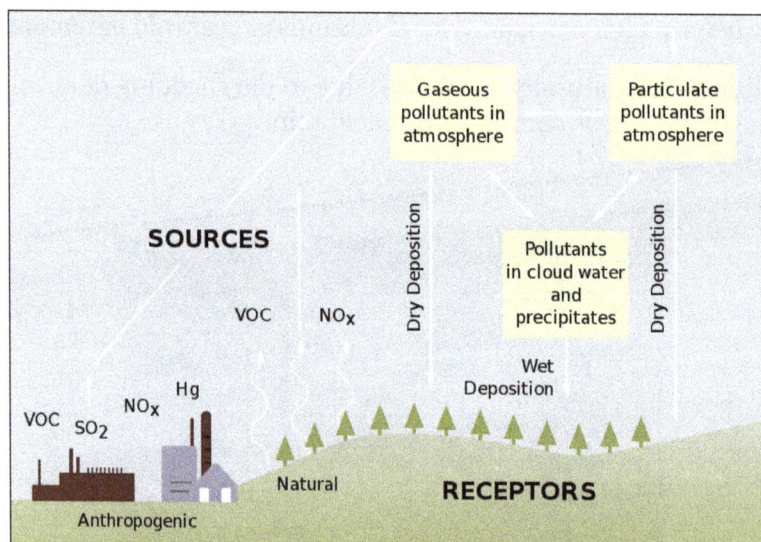

$$NO_2 + OH\cdot \rightarrow HNO_3$$

This shows the process of the air pollution being released into the atmosphere and the areas that will be affected.

## Chemistry in Cloud Droplets

When clouds are present, the loss rate of $SO_2$ is faster than can be explained by gas phase chemistry alone. This is due to reactions in the liquid water droplets.

## Hydrolysis

Sulfur dioxide dissolves in water and then, like carbon dioxide, hydrolyses in a series of equilibrium reactions:

$$SO_2 (g) + H_2O \rightleftharpoons SO_2\cdot H_2O$$

$$SO_2\cdot H_2O \rightleftharpoons H^+ + HSO_3^-$$

$$HSO_3^- \rightleftharpoons H^+ + SO_3^{2-}$$

# Oxidation

There are a large number of aqueous reactions that oxidize sulfur from S(IV) to S(VI), leading to the formation of sulfuric acid. The most important oxidation reactions are with ozone, hydrogen peroxide and oxygen (reactions with oxygen are catalyzed by iron and manganese in the cloud droplets).

# Acid Deposition

## Wet Deposition

Wet deposition of acids occurs when any form of precipitation (rain, snow, and so on.) removes acids from the atmosphere and delivers it to the Earth's surface. This can result from the deposition of acids produced in the raindrops or by the precipitation removing the acids either in clouds or below clouds. Wet removal of both gases and aerosols are both of importance for wet deposition.

## Dry Deposition

Acid deposition also occurs via dry deposition in the absence of precipitation. This can be responsible for as much as 20 to 60% of total acid deposition. This occurs when particles and gases stick to the ground, plants or other surfaces.

## Adverse Effects

Acid rain has been shown to have adverse impacts on forests, freshwaters and soils, killing insect and aquatic life-forms as well as causing damage to buildings and having impacts on human health.

## Surface Waters and Aquatic Animals

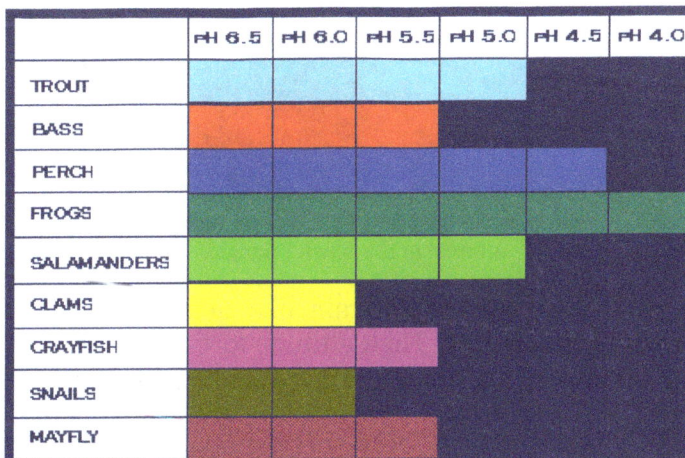

| | pH 6.5 | pH 6.0 | pH 5.5 | pH 5.0 | pH 4.5 | pH 4.0 |
|---|---|---|---|---|---|---|
| TROUT | ■ | ■ | ■ | ■ | | |
| BASS | ■ | ■ | ■ | | | |
| PERCH | ■ | ■ | ■ | ■ | ■ | |
| FROGS | ■ | ■ | ■ | ■ | ■ | ■ |
| SALAMANDERS | ■ | ■ | ■ | ■ | | |
| CLAMS | ■ | ■ | | | | |
| CRAYFISH | ■ | ■ | ■ | | | |
| SNAILS | ■ | ■ | | | | |
| MAYFLY | ■ | ■ | ■ | | | |

Not all fish, shellfish, or the insects that they eat can tolerate the same amount of acid; for example, frogs can tolerate water that is more acidic (i.e., has a lower pH) than trout.

Both the lower pH and higher aluminium concentrations in surface water that occur as a result of acid rain can cause damage to fish and other aquatic animals. At pH lower than 5 most fish eggs will not hatch and lower pH can kill adult fish. As lakes and rivers become more acidic biodiversity is reduced. Acid rain has eliminated insect life and some fish species, including the brook trout in some lakes, streams, and creeks in geographically sensitive areas, such as the Adirondack Mountains of the United States. However, the extent to which acid rain contributes directly or indirectly via runoff from the catchment to lake and river acidity (i.e., depending on characteristics of the surrounding watershed) is variable. The United States Environmental Protection Agency's (EPA) website states: "Of the lakes and streams surveyed, acid rain caused acidity in 75% of the acidic lakes and about 50% of the acidic streams". Lakes hosted by silicate basement rocks are more acidic than lakes within limestone or other basement rocks with a carbonate composition (i.e. marble) due to buffering effects by carbonate minerals, even with the same amount of acid rain.

## Soils

Soil biology and chemistry can be seriously damaged by acid rain. Some microbes are unable to tolerate changes to low pH and are killed. The enzymes of these microbes are denatured (changed in shape so they no longer function) by the acid. The hydronium ions of acid rain also mobilize toxins, such as aluminium, and leach away essential nutrients and minerals such as magnesium.

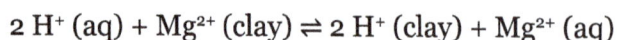

$$2\,H^+\,(aq) + Mg^{2+}\,(clay) \rightleftharpoons 2\,H^+\,(clay) + Mg^{2+}\,(aq)$$

Soil chemistry can be dramatically changed when base cations, such as calcium and magnesium, are leached by acid rain thereby affecting sensitive species, such as sugar maple (Acer saccharum).

## Forests and other Vegetation

Acid rain can have severe effects on vegetation.
A forest in the Black Triangle.

Adverse effects may be indirectly related to acid rain, like the acid's effects on soil or high concentration of gaseous precursors to acid rain. High altitude forests are especially vulnerable as they are often surrounded by clouds and fog which are more acidic than rain.

Other plants can also be damaged by acid rain, but the effect on food crops is minimized by the application of lime and fertilizers to replace lost nutrients. In cultivated areas, limestone may also be added to increase the ability of the soil to keep the pH stable, but this tactic is largely unusable in the case of wilderness lands. When calcium is leached from the needles of red spruce, these trees become less cold tolerant and exhibit winter injury and even death.

## Ocean Acidification

Acid rain has a much less harmful effect on the oceans. Acid rain can cause the ocean's pH to fall, making it more difficult for different coastal species to create their exoskeletons that they need to survive. These coastal species link together as part of the ocean's food chain and without them being a source for other marine life to feed off of more marine life will die.

Coral's limestone skeletal is sensitive to pH drop, because the calcium carbonate, core component of the limestone dissolves in acidic (low pH) solutions.

## Human Health Effects

Acid rain does not directly affect human health. The acid in the rainwater is too dilute to have direct adverse effects. The particulates responsible for acid rain (sulfur dioxide and nitrogen oxides) do have an adverse effect. Increased amounts of fine particulate matter in the air contribute to heart and lung problems including asthma and bronchitis.

## Other Adverse Effects

Effect of acid rain on statues.

Acid rain and weathering.

Acid rain can damage buildings, historic monuments, and statues, especially those made of rocks, such as limestone and marble, that contain large amounts of calcium carbonate. Acids in the rain react with the calcium compounds in the stones to create gypsum, which then flakes off.

$$CaCO_3 \text{ (s)} + H_2SO_4 \text{ (aq)} \rightleftharpoons CaSO_4 \text{ (s)} + CO_2 \text{ (g)} + H_2O \text{ (l)}$$

The effects of this are commonly seen on old gravestones, where acid rain can cause the inscriptions to become completely illegible. Acid rain also increases the corrosion rate of metals, in particular iron, steel, copper and bronze.

## Affected Areas

Places significantly impacted by acid rain around the globe include most of eastern Europe from Poland northward into Scandinavia, the eastern third of the United States, and southeastern Canada. Other affected areas include the southeastern coast of China and Taiwan.

## Prevention Methods

### Technical Solutions

Many coal-firing power stations use flue-gas desulfurization (FGD) to remove sulfur-containing gases from their stack gases. For a typical coal-fired power station, FGD will remove 95% or more of the $SO_2$ in the flue gases. An example of FGD is the wet scrubber which is commonly used. A wet scrubber is basically a reaction tower equipped with a fan that extracts hot smoke stack gases from a power plant into the tower. Lime or limestone in slurry form is also injected into the tower to mix with the stack gases and combine with the sulfur dioxide present. The calcium carbonate of the limestone produces pH-neutral calcium sulfate that is physically removed from the scrubber. That is, the scrubber turns sulfur pollution into industrial sulfates.

In some areas the sulfates are sold to chemical companies as gypsum when the purity of calcium sulfate is high. In others, they are placed in landfill. The effects of acid rain can last for generations, as the effects of pH level change can stimulate the continued leaching of undesirable chemicals into otherwise pristine water sources, killing off vulnerable insect and fish species and blocking efforts to restore native life.

Fluidized bed combustion also reduces the amount of sulfur emitted by power production. Vehicle emissions control reduces emissions of nitrogen oxides from motor vehicles.

### International Treaties

International treaties on the long-range transport of atmospheric pollutants have been agreed for example, the 1985 Helsinki Protocol on the Reduction of Sulphur Emissions under the Convention on Long-Range Transboundary Air Pollution. Canada and the US signed the Air Quality Agreement in 1991. Most European countries and Canada have signed the treaties.

### Emissions Trading

In this regulatory scheme, every current polluting facility is given or may purchase on an open market an emissions allowance for each unit of a designated pollutant it emits. Operators can then install pollution control equipment, and sell portions of their emissions allowances they no longer need for their own operations, thereby recovering some of the capital cost of their investment in such equipment. The intention is to give operators economic incentives to install pollution controls.

The first emissions trading market was established in the United States by enactment of the Clean Air Act Amendments of 1990. The overall goal of the Acid Rain Program established by the Act is to achieve significant environmental and public health benefits through reductions in emissions of sulfur dioxide ($SO_2$) and nitrogen oxides ($NO_x$), the primary causes of acid rain. To achieve this goal at the lowest cost to society, the program employs both regulatory and market based approaches for controlling air pollution.

# Arctic Haze

Arctic haze is the phenomenon of a visible reddish-brown springtime haze in the atmosphere at high latitudes in the Arctic due to anthropogenic air pollution. A major distinguishing factor of Arctic haze is the ability of its chemical ingredients to persist in the atmosphere for an extended period of time compared to other pollutants. Due to limited amounts of snow, rain, or turbulent air to displace pollutants from the polar air mass in spring, Arctic haze can linger for more than a month in the northern atmosphere.

Arctic haze was first noticed in 1750 when the Industrial Revolution began. Explorers and whalers could not figure out where the foggy layer was coming from. *"Poo-jok"* was the term the Inuit used for it. Another hint towards clarifying this issue was relayed in notes approximately a century ago by Norwegian explorer Fridtjof Nansen. After trekking through the Arctic he found dark stains on the ice. The term "Arctic haze" was coined in 1956 by J. Murray Mitchell, a US Air Force officer stationed in Alaska, to describe an unusual reduction in visibility observed by North American weather reconnaissance planes. From his investigations, Mitchell thought the haze had come from industrial areas in Europe and China. He went on to become an eminent climatologist. The haze is seasonal, reaching a peak in late winter and spring. When an aircraft is within a layer of Arctic haze, pilots report that horizontal visibility can drop to one tenth that of normally clear sky. At this time it was unknown whether the haze was natural or was formed by pollutants.

In 1972, Glenn Shaw of the Geophysical Institute at the University of Alaska attributed this smog to transboundary anthropogenic pollution, whereby the Arctic is the recipient of contaminants whose sources are thousands of miles away. Further research continues with the aim of understanding the impact of this pollution on global warming.

The pollutants are commonly thought to originate from coal-burning in northern mid-latitudes, especially in Asia. The aerosols contain about 90% sulfur and the rest is carbon, which makes the haze reddish in color. This pollution is helping the Arctic warm up faster than any other region, although increases in greenhouse gases are the main driver of this climatic change.

Sulfur aerosols in the atmosphere affect cloud formation, leading to localized cooling effects over industrialized regions due to increased reflection of sunlight, which masks the opposite effect of trapped warmth beneath the cloud cover. During the Arctic winter, however, there is no sunlight to reflect. In the absence of this cooling effect, the dominant effect of changes to Arctic clouds is an increased trapping of infrared radiation from the surface.

Ship emissions, mercury, aluminium, vanadium, manganese, and aerosol and ozone pollutants are many examples of the pollution that is affecting this atmosphere, but the smoke from forest fires is not a significant contributor. Some of those pollutants figure among environmental effects of coal burning. Due to low deposition rates, these pollutants are not yet having adverse effects on people or animals. Different pollutants actually represent different colors of haze. Dr. Shaw discovered in 1976 that the yellowish haze is from dust storms in China and Mongolia. The particles were carried polewards by unusual air currents. The trapped particles were dark gray the next year he took a sample. That was caused by a heavy amount of industrial pollutants.

# Impacts of Air Pollution and Acid Rain on Vegetation

Sulphur dioxide and nitrogen oxides both combine with water in the atmosphere to create acid rain. Acid rain acidifies the soils and waters where it falls, killing off plants. Many industrial processes produce large quantities of pollutants including sulphur dioxide and nitrous oxide. These are also produced by car engines and are emitted in the exhaust. When sulphur dioxide and nitrous oxide react with water vapour in the atmosphere, acids are produced. The result is what is termed acid rain, which causes serious damage to plants.

In addition, other gaseous pollutants, such as ozone, can also harm vegetation directly.

## How Acid Rain Harms Trees

Acid rain does not usually kill trees directly. Instead, it is more likely to weaken the trees by damaging their leaves, limiting the nutrients available to them, or poisoning them with toxic substances slowly released from the soil. The main atmospheric pollutants that affect trees are nitrates and sulphates. Forest decline is often the first sign that trees are in trouble due to air pollution.

Scientists believe that acidic water dissolves the nutrients and helpful minerals in the soil and then washes them away before the trees and other plants can use them to grow. At the same time, the acid rain causes the release of toxic substances such as aluminium into the soil. These are very harmful to trees and plants, even if contact is limited. Toxic substances also wash away in the run-off that carries the substances into streams, rivers, and lakes. Fewer of these toxic substances are released when the rainfall is cleaner.

Even if the soil is well buffered, there can be damage from acid rain. Forests in high mountain regions receive additional acid from the acidic clouds and fog that often surround them. These clouds and fog are often more acidic than rainfall. When leaves are frequently bathed in this acid fog, their protective waxy coating can wear away. The loss of the coating damages the leaves and creates brown spots. Leaves turn the energy in sunlight into food for growth. This process is called photosynthesis. When leaves are damaged, they cannot produce enough food energy for the tree to remain healthy.

Once trees are weak, diseases or insects that ultimately kill them can more easily attack them. Weakened trees may also become injured more easily by cold weather.

## How Air Pollution Harms Trees

Whilst acid rain is a major cause of damage to vegetation, air pollutants which can also be harmful directly. These include sulphur dioxide and ozone.

Sulphur Dioxide: Sulphur dioxide, one of the main components of acid rain, has direct effects on vegetation. Changes in the physical appearance of vegetation are an indication that the plants' metabolism is impaired by the concentration of sulphur dioxide. Harm caused by sulphur dioxide is first noticeable on the leaves of the plants. For some plants injury can occur within hours or days of being exposed to high levels of sulphur dioxide. It is the leaves in mid-growth that are the most vulnerable, while the older and younger leaves are more resistant. You can see the damage to coniferous needles by observing the extreme colour difference between the green base and the bright orange-red tips.

The effects of sulphur dioxide are influenced by other biological and environmental factors such as plant type, age, sunlight levels, temperature, humidity and the presence of other pollutants (ozone and nitrogen oxides). Thus, even though sulphur dioxide levels may be extremely high, the levels may not affect vegetation because of the surrounding environmental conditions. It is also possible that the plants and soils may temporarily store pollutants. By storing the pollutants they are preventing the pollutants from reacting with other substances in the plants or soil.

Ozone: The effects of ozone on plants have been investigated intensively for almost two decades. Studies made in controlled environment (CE) chambers, glasshouses and in the field, using open-topped chambers, have all contributed to the understanding of the mechanisms underlying ozone effects and their ultimate impact on vegetation. The biochemical mechanisms by which ozone interacts with plants have been intensively studied and, although the relative significance of different initial reactions remains unclear, there is a consensus that the key event in plant responses is oxidative damage to cell membranes. This primary oxidative damage results in the loss of membrane integrity and function, and in turn to inhibition of essential biochemical and physiological processes. A key target is photosynthesis, although ozone may also affect stomatal function and so modify plant responses to other factors, such as drought and elevated carbon dioxide. These changes result in reduced growth and yield in many plants. However, it is clear that such responses vary in magnitude between species and also between different cultivars within species. The mechanisms by which some species and genotypes are protected from ozone injury are not clear, but may include differences in uptake into the leaf or in the various components of antioxidant metabolism. Ozone may also increase the severity of many fungal diseases, while virus infections reduce the effects of ozone in some plants.

Acid deposition and ozone exposure have increased considerably in the past 50 years in Asia, Europe and the US, with many reports of tree/forest decline and increased mortality. In general, the more highly polluted forests have the higher rate of decline and mortality. However, there has been no recent chronic deterioration in the UK of tree condition. Since the early 1990s, peak concentrations of ozone have been falling, whilst the large reduction in sulphur dioxide emissions since the 1970s has provided an opportunity for recovery of many plant species.

## Acidification by Forestry

While forestry has long been considered to be adversely affected by air pollution and acid rain, recent studies show it to be part of the acidifying process. The rough canopies of mature evergreen forests are efficient scavengers of particulate and gaseous contaminants in polluted air. This results in a more acidic deposition under the forest canopies than in open land. Chemical processes at the roots of trees, evergreens in particular, further acidify the soil and soil water in forest catchments. When the forests are located on poorly buffered soils, these processes can lead to a significant acidification of the run-off water and consequent damage to associated streams and lakes.

## Effects of Air Pollution on Agricultural Crops

Agricultural crops can be injured when exposed to high concentrations of various air pollutants. Injury ranges from visible markings on the foliage, to reduced growth and yield, to premature death of the plant. The development and severity of the injury depends not only on the concentration of the particular pollutant, but also on a number of other factors. These include the length of exposure to the pollutant, the plant species and its stage of development as well as the environmental factors conducive to a build-up of the pollutant and to the preconditioning of the plant, which make it either susceptible or resistant to injury.

## Effects of Air Pollution on Plants

Air pollution injury to plants can be evident in several ways. Injury to foliage may be visible in a short time and appear as necrotic lesions (dead tissue), or it can develop slowly as a yellowing or chlorosis of the leaf. There may be a reduction in growth of various portions of a plant. Plants may be killed outright, but they usually do not succumb until they have suffered recurrent injury.

## Oxidants

Ozone is the main pollutant in the oxidant smog complex. Its effect on plants was first observed in the Los Angeles area in 1944. Since then, ozone injury to vegetation has been reported and documented in many areas throughout North America, including the southwestern and central regions of Ontario. Throughout the growing season, particularly July and August, ozone levels vary significantly. Periods of high ozone are associated with regional southerly air flows that are carried across the lower Great Lakes after passing over many urban and industrialised areas of the United States. Localized, domestic ozone levels also contribute to the already high background levels. Injury levels vary annually and white bean, which are particularly sensitive, are often used as an indicator of damage. Other sensitive species include cucumber, grape, green bean, lettuce, onion, potato, radish, rutabagas, spinach, sweet corn, tobacco and tomato. Resistant species include endive, pear and apricot.

Ozone symptoms characteristically occur on the upper surface of affected leaves and appear as a flecking, bronzing or bleaching of the leaf tissues. Although yield reductions are usually with visible foliar injury, crop loss can also occur without any sign of pollutant stress. Conversely, some crops can sustain visible foliar injury without any adverse effect on yield.

Ozone injury to soybean foliage.

Susceptibility to ozone injury is influenced by many environmental and plant growth factors. High relative humidity, optimum soil-nitrogen levels and water availability increase susceptibility. Injury development on broad leaves also is influenced by the stage of maturity. The youngest leaves are resistant. With expansion, they become successively susceptible at middle and basal portions. The leaves become resistant again at complete maturation.

## Sulfur Dioxide

Major sources of sulfur dioxide are coal-burning operations, especially those providing electric power and space heating. Sulfur dioxide emissions can also result from the burning of petroleum and the smelting of sulfur containing ores.

Sulfur dioxide enters the leaves mainly through the stomata (microscopic openings) and the resultant injury is classified as either acute or chronic. Acute injury is caused by absorption of high concentrations of sulfur dioxide in a relatively short time. The symptoms appear as 2-sided (bifacial) lesions that usually occur between the veins and occasionally along the margins of the leaves. The colour of the necrotic area can vary from a light tan or near white to an orange-red or brown depending on the time of year, the plant species affected and weather conditions. Recently expanded leaves usually are the most sensitive to acute sulfur dioxide injury, the very youngest and oldest being somewhat more resistant.

Acute sulfur dioxide injury to raspberry.

Chronic injury is caused by long-term absorption of sulfur dioxide at sub-lethal concentrations. The symptoms appear as a yellowing or chlorosis of the leaf, and occasionally as a bronzing on the under surface of the leaves.

Different plant species and varieties and even individuals of the same species may vary considerably in their sensitivity to sulfur dioxide. These variations occur because of the differences in geographical location, climate, stage of growth and maturation. The following crop plants are generally considered susceptible to sulfur dioxide: alfalfa, barley, buckwheat, clover, oats, pumpkin, radish, rhubarb, spinach, squash, Swiss chard and tobacco. Resistant crop plants include asparagus, cabbage, celery, corn, onion and potato.

## Fluoride

Fluorides are discharged into the atmosphere from the combustion of coal; the production of brick, tile, enamel frit, ceramics, and glass; the manufacture of aluminium and steel; and the production of hydrofluoric acid, phosphate chemicals and fertilizers.

Fluorides absorbed by leaves are conducted towards the margins of broad leaves (grapes) and to the tips of monocotyledonous leaves (gladiolus). Little injury takes place at the site of absorption, whereas the margins or the tips of the leaves build up injurious concentrations. The injury starts as a gray or light-green water-soaked lesion, which turns tan to reddish-brown. With continued exposure the necrotic areas increase in size, spreading inward to the midrib on broad leaves and downward on monocotyledonous leaves.

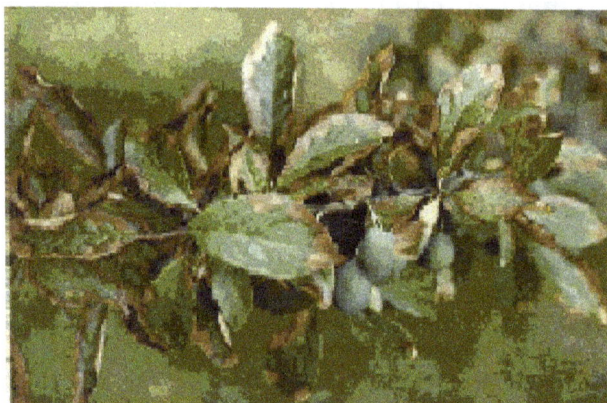

Fluoride injury to plum foliage. The fluoride enters the leaf through the stomata and is moved to the margins where it accumulates and causes tissue injury. The characteristic dark band separating the healthy (green) and injured (brown) tissues of affected leaves.

Studies of susceptibility of plant species to fluorides show that apricot, barley (young), blueberry, peach (fruit), gladiolus, grape, plum, prune, sweet corn and tulip are most sensitive. Resistant plants include alfalfa, asparagus, bean (snap), cabbage, carrot, cauliflower, celery, cucumber, eggplant, pea, pear, pepper, potato, squash, tobacco and wheat.

## Ammonia

Severe ammonia injury to apple foliage and subsequent recovery through the production of new leaves following the fumigation.

Ammonia injury to vegetation has been observed frequently in Ontario in recent years following accidents involving the storage, transportation or application of anhydrous and aqua ammonia fertilizers. These episodes usually release large quantities of ammonia into the atmosphere for brief periods of time and cause severe injury to vegetation in the immediate vicinity.

Complete system expression on affected vegetation usually takes several days to develop, and appears as irregular, bleached, bifacial, necrotic lesions. Grasses often show reddish, interveinal necrotic streaking or dark upper surface discolouration. Flowers, fruit and woody tissues usually are not affected, and in the case of severe injury to fruit trees, recovery through the production of new leaves can occur. Sensitive species include apple, barley, beans, clover, radish, raspberry and soybean. Resistant species include alfalfa, beet, carrot, corn, cucumber, eggplant, onion, peach, rhubarb and tomato.

## Particulate Matter

Particulate matter such as cement dust, magnesium-lime dust and carbon soot deposited on vegetation can inhibit the normal respiration and photosynthesis mechanisms within the leaf. Cement dust may cause chlorosis and death of leaf tissue by the combination of a thick crust and alkaline toxicity produced in wet weather. The dust coating also may affect the normal action of pesticides and other agricultural chemicals applied as sprays to foliage. In addition, accumulation of alkaline dusts in the soil can increase soil pH to levels adverse to crop growth.

Cement-dust coating on apple leaves and fruit. The dust had no injurious effect on the foliage, but inhibited the action of a pre-harvest crop spray.

## Investigation of Air Pollution Injury to Vegetation

The Ministry of the Environment monitors air quality at 33 stations across the province. The sites are set in both urban and rural settings and monitor the 6 most common air pollutants: sulfur dioxide, ozone, nitrogen dioxide, total reduced sulfur compounds, carbon monoxide and suspended particles. The sites are monitored in real-time on an hourly basis. Pollutant concentrations are converted into an Air Quality Index (AQI) with a lower AQI translating into cleaner air. AQI values above 50 can cause crop injury.

# Economic Effects of Air Pollution

There are numerous effects of air pollution on the ecosystem which in turn have various economic implications. In simple terminology, we can say that air pollution effects can be both direct and indirect. For instance, pollution of air primarily causes respiratory and other health hazards in people who are being directly exposed to various harmful gases. The secondary, and long run impact, would be that following the health problems, the productivity of workers might be adversely affected which in turn hamper output levels. This is how air pollution exerts an indirect effect on the overall economy.

Information on physical and biological responses, including human health, to alternative environmental states forms the traditional scientific basis for setting environmental standards. Estimates of the benefits and costs associated with environmental regulation can be used in support of such standards by providing an additional measure of efficiency of environmental management programs. With respect to air pollution control, the Clean Air Act in foreign specifically states that regulatory decisions are to be based on human health and other physical and biological science information, not economic efficiency criteria.

Some forms of air pollution are generated primarily by anthropogenic sources which is the term used for humans. Others, such as ozone, occur both naturally and anthropogenically. The recognition of anthropogenic sources of pollutants and their biological and economic effects on managed ecosystems such as agricultural crops and forests provided impetus for air pollution control programs.

The link between Economics and natural science economic analysis can be an effective tool for comparing the costs and benefits of alternative resource or environmental management policy actions. When correctly formulated, such economic analyses can be useful in estimating the monetary values of vegetation and other receptor losses from air pollution or the welfare consequences of air pollution reductions. Economic assessments of air pollution damages typically start with biological information on observed or inferred vegetation quality and/or yield losses for alternative air quality levels. The economic loss or benefit associated with that biological response is then calculated based on the behavior of economic agents as reflected in market information (prices).

Pollutants which are, or have a potential to be, harmful to vegetation include oxides of sulfur, oxides of nitrogen, particulates and secondary products such as photochemical oxidants and acid deposition. Sulfur oxides are believed to be the primary precursor of acid rainfall and are primarily emitted from stationary sources such as utility and industrial boilers burning coal as a fuel. Both stationary and transportation-related sources, such as natural gas refineries, cars and trucks, contribute to nitrogen oxides emissions. Empirical evidence shows that anthropogenic emissions of sulfur and nitrogen oxides in the eastern United States increased significantly between 1950 and 1970. Since 1970, however, sulfur oxide emissions have declined while nitrogen oxides have continued to rise.

The cost of air pollution to the world's most advanced economies plus India and China is estimated to be US $ 3.5 trillion per year in lives lost and ill health.

In OECD countries the monetary impact of death and illness due to outdoor air pollution in 2010 is estimated to have been US $ 1.7 trillion.

There is an urgent need to reduce levels of air pollution globally. Although air quality measures have had positive results at some locations in the world, millions of people in both developing and developed countries die prematurely every year because of long-term exposure to air pollutants.

The health of many more is seriously affected which in turn drags other crucial issues those affect the economy. Most cities where outdoor air pollution is monitored do not meet the World Health Organization (WHO) guidelines for acceptable pollutant levels. People who live in these cities have increased risks of stroke, heart disease, lung cancer, chronic and acute respiratory diseases (including asthma) and other health problems. Indoor air pollution is another major cause of poor health and premature death, especially in developing countries.

Sources of air pollution include traffic (especially diesel vehicles), industrial sectors (from brick making to oil and gas production), power plants, cooking and heating with solid fuels (e.g. coal, wood, crop waste), forest fires and open burning of municipal waste and agricultural residues.

The health impacts of air pollution are much larger than was thought only a few years ago. The WHO estimated that in 2012 around 7 million premature deaths resulted from air pollution, more than double previous estimates. The new estimate is based on increasing knowledge of air pollution-related diseases and use of improved air quality measurements and technology. According to WHO, outdoor air pollution caused 3.7 million premature deaths in 2012. Indoor air pollution is responsible for about 4.3 million premature deaths every year.

Outdoor air quality is rapidly deteriorating in major cities in low and middle income countries (LMICs). The WHO guideline for average annual particulate matter ($PM_{10}$) levels is 20 micrograms per cubic meter. For fine particles ($PM_{2.5}$) the average annual level is 10 micrograms per cubic metres and 25 micrograms per cubic meter for a 24 hour period.

Air pollution levels in cities in LMICs sometimes far exceed these levels. About 3 billion people in the world cook and heat their homes with coal and biomass. Air pollutant emissions need to be reduced not only from these inefficient energy systems, but also from agricultural waste incineration, forest fires and charcoal production. In Africa, due to the rapid growth of its cities and megacities, a large increase in air pollutant emissions from burning of fossil fuels and traditional biomass use for energy services is expected in the near future.

Underscoring the fact that air pollution exerts adverse economic effects that cripple economic growth attainment, it is extremely important to adopt effective restrictive measures to mitigate air pollution. It is not only that the economy of the country that is responsible for pollution is being affected solely. Rather, it's a global phenomenon and this justifies impositions of both domestic and international clean air acts.

## References

- Weathers, K. C. And Likens, G. E. (2006). "Acid rain", pp. 1549–1561 in: W. N. Rom and S. Markowitz (eds.). Environmental and Occupational Medicine. Lippincott-Raven Publ., Philadelphia. Fourth Edition, ISBN 0-7817-6299-5

- Health-impacts-air-pollution, health: edf.org, Retrieved 31 July, 2019

- Leung, L. Ruby; Gustafson Jr, William I. (Aug 26, 2005). "Potential regional climate change and implications to U.S. air quality". Geophysical Research Letters. 32 (16). Bibcode:2005georl..3216711L. Doi:10.1029/2005GL022911

- Kesler, Stephen (2015). Mineral Resources, Economics and the Environment. Cambridge University. ISBN 9781107074910

Air pollution-and-brain, fact-sheets, resources: arb.ca.gov, Retrieved 17 February , 2019

- Leung, L. Ruby; Gustafson Jr, William I. (Aug 26, 2005). "Potential regional climate change and implications to U.S. air quality". Geophysical Research Letters. 32 (16). Bibcode:2005georl..3216711L. Doi:10.1029/2005GL022911

- Robert Tardif (2002). Fog characteristics. Archived 2011-05-20 at the Wayback Machine University Corporation for Atmospheric Research. Retrieved on 2007-02-11

- Indoor-air-quality-health-effects: bluepointenvironmental.com , Retrieved 25 August, 2019

# 4

# Air Quality Monitoring, Measurement and Health Index

Air quality monitoring refers to the measurement of air quality using various techniques such as air sampling. Air quality health index is a scale which is used to analyze the impact which air quality has on health. The diverse aspects of air quality monitoring as well as air quality health index have been thoroughly discussed in this chapter.

## Air Quality Measurement

There are many ways to measure air pollution, with both simple chemical and physical methods and with more sophisticated electronic techniques. There are four main methods of measuring air pollution.

Passive sampling methods provide reliable, cost-effective air quality analysis, which gives a good indication of average pollution concentrations over a period of weeks or months. Passive samplers are so-called because the device does not involve any pumping. Instead the flow of air is controlled by a physical process, such as diffusion. Diffusion tubes are simple passive samplers, which provide very useful information regarding ambient air quality. They are available for a number of pollutants, but are most commonly and reliably used for nitrogen dioxide and benzene. The tubes, which are 71mm long with an internal diameter of 11mm, contain two stainless steel gauzes placed at one end of a short cylinder. The steel gauzes contain a coating of triethanolamine, which converts the nitrogen dioxide to nitrite. The accumulating nitrates are trapped within the steel gauze, ready for laboratory analysis. The tube is open to the atmosphere at the other end, which is exposed downwards to prevent rain or dust from entering the tube. To ensure that all the nitrogen dioxide originates from the test site, the tubes are sealed before and after exposure. The tubes are manually distributed and collected, and are analysed in a laboratory.

Active sampling methods use physical or chemical methods to collect polluted air, and analysis is carried out later in the laboratory. Typically, a known volume of air is pumped through a collector (such as a filter, or a chemical solution) for a known period of time. The collector is later removed for analysis. Samples can be collected daily, providing measurements for short time periods, but at a lower cost than automatic monitoring methods.

Automatic methods produce high-resolution measurements of hourly pollutant concentrations or better, at a single point. Pollutants analysed include ozone, nitrogen oxides, sulphur dioxide, carbon monoxide and particulates. The samples are analysed using a variety of methods including spectroscopy and gas. The sample, once analysed is downloaded in real-time, providing very accurate information.

Remote optical/long path-analysers use spectroscopic techniques, make real-time measurements of the concentrations of a range of pollutants including nitrogen dioxide and sulphur dioxide.

The amount of pollution in the air, however sampled, is usually measured by its concentration in air. The concentration of a pollutant in air may be defined in terms of the proportion of the total volume that it accounts for. Concentrations of pollutant gases in the atmosphere are usually measured in parts per million by volume (ppmv), parts per billion by volume (ppbv) or parts per trillion (million million) by volume (pptv). Pollutant concentrations are also measured by the weight of pollutant within a standard volume of air, for example microgrammes per cubic metre ($\mu gm^{-3}$) or milligrammes per cubic metre ($mgm^{-3}$).

# Air Pollution Monitoring System

Air pollution emerged in many parts of the world as a result of explosive industrial growth. Road transport is also one of the major contributors of air pollution which contribute to climate change that has perilous domestic and global consequences.

Generation and transport of pollutant materials are governed not only by the distributions of their sources but also by the dynamics of the atmosphere. Pollutant clouds are sometimes observed traveling along the wind directions.

To understand the involved processes in more detail we need more thorough data on the spreads of fine-grain pollutants and their variations with time. An air pollution monitoring system that is comprehensive in terms of spatial and pollutant coverage and is relatively inexpensive and autonomous is the priority.

Some of the existing instruments for air pollution monitoring are Fourier transform infrared (FTIR) instruments, gas chromatographs and mass spectrometers. These instruments provide fairly accurate and selective gas readings. A gas sensor that is compact, robust with versatile applications and low cost could be an equally effective alternative. Some of the gases monitoring technologies are electrochemical, infrared, catalytic bead, photo ionization and solid-state. The existing monitoring system largely uses smart transducer interface module (STIM) with semiconductor gas sensors which uses the 1451.2 standard.

STIM was found to an efficient monitoring system but for the power requirements and ability to expand for large deployment. One of the large scale sensor networks for monitoring and forecasting is Environment Observation and Forecasting System (EOFS). Air pollution monitoring system based on geo sensor network with control action and adaptive sampling rates proposed in also cannot be vast deployment due to high cost.

Now in this project we are using locally available gas sensor for observing the polluted gases like Carbon monoxide (CO), Carbon dioxide ($CO_2$) and parameters like temperature, humidity. By using this method people can view the level of pollution through wireless system. It reduced cost, reliable and comfortable for any place where we are monitoring the gases.

## Toxic Gases

Toxic gases are carbon dioxide and carbon monoxide. These gases are very harmful and dangerous to the people.

## Carbon Monoxide (CO)

### Nature and Sources of the Pollutant

Carbon monoxide is a colorless, odorless and poisonous gas formed when carbon in fuels is not burned completely. It is a byproduct of highway vehicle exhaust, which contributes about 60 percent of all CO emissions nationwide. In cities automobile exhaust can cause as much as 95 percent of all CO emissions.These emissions can result in high concentrations of CO particularly in local areas with heavy traffic congestion. Other sources of CO emissions include industrial processes and fuel combustion in sources such as boilers and incinerators. Despite an overall downward trend in concentrations and emissions of CO some metropolitan areas still experience high levels of CO.

### Health and Environmental Effects

Carbon monoxide enters the bloodstream and reduces oxygen delivery to the body's organs and tissues. The health threat from exposure to CO is most serious for those who suffer from cardiovascular disease. Healthy individuals are also affected but only at higher levels of exposure. Exposure to elevated CO levels is associated with visual impairment, reduced work capacity, reduced manual dexterity, poor learning ability and difficulty in performing complex tasks.

Environmental Protection Agency (EPA)'s health-based national air quality standard for CO is 9 parts per million (ppm) measured as an annual second-maximum 8-hour average concentration.

### Trends in Carbon Monoxide Level

Long-term improvements continued between 1986 and 1995. National average CO concentrations decreased 37 percent while CO emissions decreased 16 percent. Long-term air quality improvement in CO occurred despite a 31 percent increase in vehicle miles traveled in the U.S. during the past 10 years.

Between 1994 and 1995, national average CO concentrations decreased 10 percent, while total CO emissions decreased 7 percent. Transportation sources (includes highway and off-highway vehicles) now account for 81 percent of national total CO emissions.

## CO Concentration

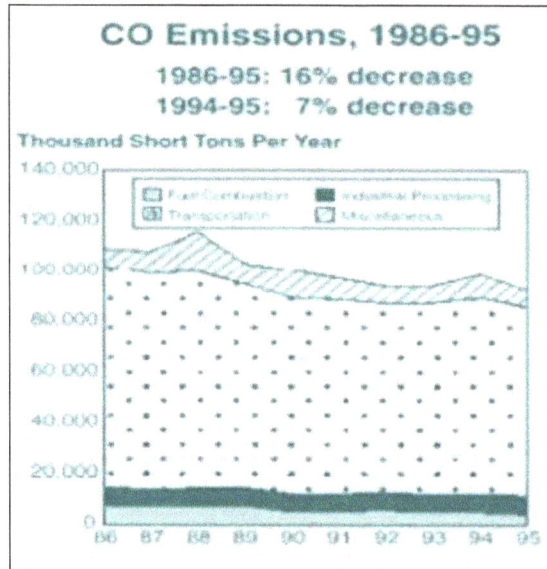

Concentration of CO.

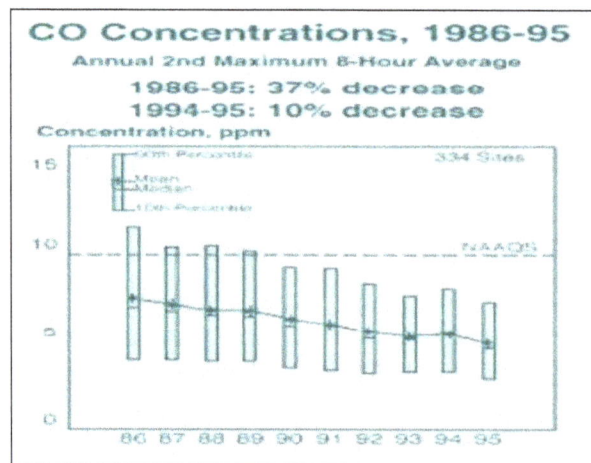

Concentration of CO (8 hour average).

## Carbon Dioxide (CO$_2$)

Carbon dioxide (CO$_2$) is a colorless, odorless and non-flammable gas that is a product of cellular respiration and burning of fossil fuels. It has a molecular weight of 44.01g/mol. Although it is typically present as a gas carbon dioxide also can be a solid form as dry ice and liquefied depending on temperature and pressure.

This gas is utilized by many types of industry including breweries, mining ore, manufacturing of carbonated drinks, drugs, disinfectants, pottery and baking powder. It also is a primary gas associated with volcanic eruptions. $CO_2$ acts to displace oxygen, making compressed $CO_2$ the main ingredient in fire extinguishers.

Occupations that are most at risk from $CO_2$ exposure include miners, brewers, carbonated beverage workers and grain elevator workers. $CO_2$ is present in the atmosphere at 0.035%. In terms of worker safety, Occupational Safety and Health Administration (OSHA) has set a permissible exposure limit (PEL) for $CO_2$ of 5,000 parts per million (ppm) over an 8-hour work day, which is equivalent to 0.5% by volume of air.Similarly the American Conference of Governmental Industrial Hygienists (ACGIH) TLV (threshold limit value) is 5,000 ppm for an 8hour workday with a ceiling exposure limit of 30,000 ppm for a 10-minute period based on acute inhalation data.

A value of 40,000 ppm is considered immediately dangerous to life and health based on the fact that a 30-minute exposure to 50,000 ppm produces intoxication and concentrations greater than that (7-10%) produce unconsciousness. Additionally acute toxicity data show the lethal concentration low for $CO_2$ is 90,000 ppm (9%) over 5 minutes. See table for a listing of regulatory agency standards for acceptable $CO_2$ concentrations in the workplace. $CO_2$ is a good indicator of proper building ventilation and indoor air exchange rates. Consequently it is measured in buildings to determine if the indoor air is adequate for humans to occupy the building.

## Symptoms from Low to High Concentrations of $CO_2$

Table: Tabulation for toxic gas $CO_2$.

| %$CO_2$ | Symptoms |
|---|---|
| 2 to 3 | Shortness of breath, deep breathing. |
| 5 | Breathing becomes heavy, sweating, pulse quickens. |
| 7.5 | Headaches, dizziness, restlessness, breathlessness, Increased heart rate and blood pressure, visual distortion. |
| 10 | Impaired hearing, nausea, vomiting, loss of consciousness. |
| 30 | Coma, convulsions, death. |

## Properties

Carbon dioxide ($CO_2$) is a colorless and odorless gas. It is non-flammable and chemically non-reactive. $CO_2$ is 1.5 times as heavy as air (its density is 1.80 g L-1 at 25 °C and 1 atm) and if it is emitted slowly, flows down-slope and may accumulate at low elevations. Concentration ranges of $CO_2$ in dilute volcanic plumes can range from 1 ppm to hundreds of ppm above the troposphere background of ~360 ppm and the gas has a residence times in the lower atmosphere of approximately 4 years. Due to the high levels of $CO_2$ required to cause harm, concentrations of $CO_2$ are often expressed as a percentage of the gas in air by volume (1% = 10,000 ppm). This is in contrast to other volcanic gases.

## Hardware Description

### Block Diagram

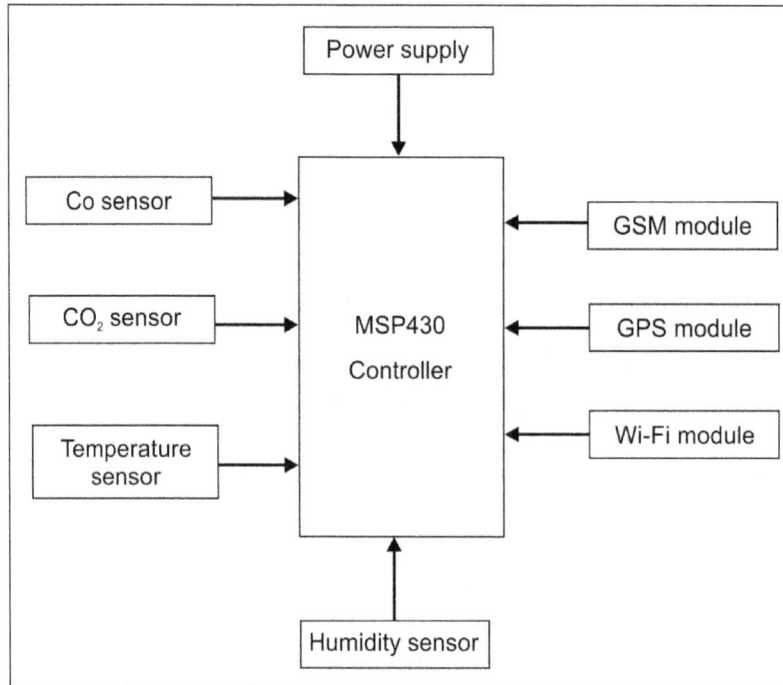

The block diagram consists of:

- Power supply
- Sensor network
- Connector launch pad
- Internet part

### Power Supply

This introduces the operation of power supply circuits built using filters, rectifiers and then voltage regulators. Starting with an AC voltage, a steady DC voltage is obtained by rectifying the AC voltage, then filtering to a DC level, and finally regulating to obtain a desired fixed DC voltage. The regulation is usually obtained from an IC voltage regulator unit, which remain the same if the input DC voltage varies or the output load connected to DC voltage changes.

A block diagram containing the parts of a typical power supply is shown below. The AC voltage, typically 120 Vrms is connected to transformer which steps that AC voltage down to the level for the desired DC output.

$$\text{AC input} \rightarrow \text{step down transformer} \rightarrow \text{rectifier} \rightarrow \text{filter} \rightarrow \text{regulator}$$

A diode rectifier that provides a full-wave rectified voltage that is initially filtered by a simple capacitor filter to produce a DC voltage. A regulated circuit can use this DC inputs to provide a DC voltage that not only has much less ripple voltage but also remains the same DC value even if the input DC voltage varies somewhat or the load connected to the output DC voltage changes this voltage regulation is usually obtained using one of a number of popular voltage regulation IC unit.

Power supply consists of following unit:

- Step down transformer
- Rectifier unit
- Input filter
- Regulator unit

## Circuit Diagram

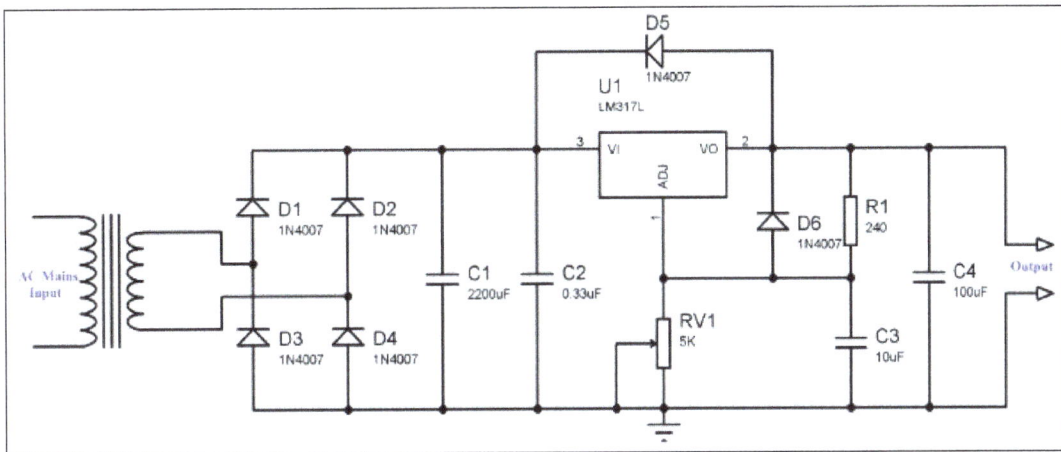

Circuit diagram of power supply.

## Step-down Transformer

It is used to step down the main supply voltage by using step down transformer. It consists of primary and secondary coils. The output from the secondary is also AC wave form. so we have to convert AC voltage in to DC voltage by using rectifier unit.

## Full Wave Rectifier

The full wave rectifier circuit is one that is widely used for power supplies and many other areas where a full wave rectification is required. Full wave rectification can also be achieved using a bridge rectifier which is made of four diodes. The full wave rectifier circuit is used in most rectifier applications because of the advantages it offers. It can provide better rectification than other rectifiers. So we have desired to use full wave rectifier for rectification.

## Advantages

- More efficient use of the transformer.
- Utilizes both halves of the AC wave form.
- Easier to provide smoothing as a result of ripple frequency.

## Filter

If a capacitor is added in parallel with the load resistor of a rectifier to form a simple Filter Circuit, the output of the rectifier will be transformed into a more stable DC Voltage. The principle of the capacitor is charging and discharging. At first, the capacitor is charged to the peak value of the rectified waveform.

Beyond the peak, the capacitor is discharged through the load resistor until the time at which the rectified voltage exceeds the capacitor voltage. Then the capacitor is charged again and the process repeats itself. There are two types of filters used in the power supply circuit. They are:

- Input filter: If the capacitor is added before the regulator it will act as an input filter. It is used to filter the ripples from rectified output.

- Output filter: If the capacitor is added after the regulator it will act as an output filter. Even though the regulated output does not have many ripples again the IC regulators output is given to the output filter then to the load.

## Ic Voltage Regulator

The MC78XX/LM78XX/MC78XXA series of three terminal positive regulators are available with several fixed output voltages, making them useful in a wide range of applications. Each type employs internal current limiting, thermal shut down and safe operating area protection, making it essentially indestructible. If adequate heat sinking is provided, they can deliver over 1A output current. Although designed primarily as fixed voltage regulators, these devices can be used with external components to obtain adjustable voltages and currents.

A regulated power supply is very much essential for several electronic devices due to the semiconductor material employed in them have a fixed rate of current as well as voltage. The device may get damaged if there is any deviation from the fixed rate. The AC power supply gets converted into constant DC by this circuit. By the help of a voltage regulator DC, unregulated output will be fixed to a constant voltage. The circuit is made up of linear voltage regulator 7805 along with capacitors and resistors with bridge rectifier made up from diodes.

## Features

- Output current up to 1a

- Output voltages of 5, 6, 8, 9, 10, 12, 15, 18, 24v

- Thermal overload protection

- Short circuit protection

- Output transistor safe operating area protection

## Sensor Network

- Temperature sensor

- Humidity sensor

- CO sensor

- $CO_2$ sensor

## Temperature Sensor

Temperature sensors are vital to a variety of everyday products. For example, household ovens, refrigerators and thermostats all rely on temperature maintenance and control in order to function properly. Temperature control also has applications in chemical engineering. Examples of this include maintaining the temperature of chemical reactor at ideal set-point, monitoring the temperature of possible runaway reaction to ensure the safety of employees and maintaining the temperature of streams released to the environment to minimize harmful environmental impact.

## LM35 Temperature Sensor

The LM35 series are precision integrated-circuit temperature devices with an output voltage linearlyproportional to the Centigrade temperature. The LM35 device has an advantage over linear temperature sensors calibrated in Kelvin as the user is not required to subtract a large constant voltage from the output to obtain convenient Centigrade scaling.

The LM35 device does not require any external calibration or trimming to provide typical accuracies of ± ¼ °C at room temperature and ± ¾ °C over a full −55 °C to 150 °C temperature range. Lower cost is assured by trimming and calibration at the wafer level. The low-output impedance, linear output and precise inherent calibration of the LM35 device makes interfacing to readout or control circuitry especially easy.

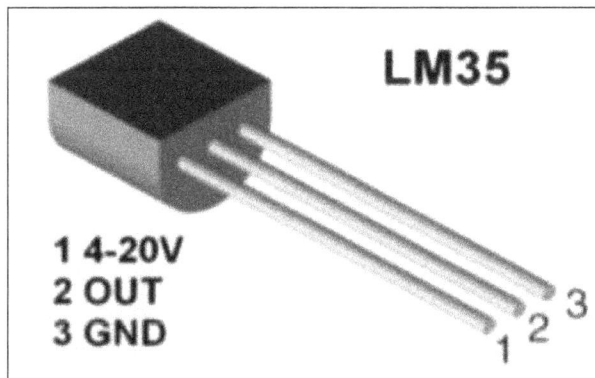

Temperature sensor LM35.

The device is used with single power supplies or with plus and minus supplies. As the LM35 device draws only 60 µA from the supply, it has very low self-heating of less than 0.1 °C in still air. The LM35 device is rated to operate over a −55 °C to 150 °C temperature range, while the LM35C device is rated for a −40 °C to 110 °C range (−10° with improved accuracy).

The LM35-series devices are available packaged in hermetic to transistor packages, while the LM35C, LM35CA and LM35D devices are available in the plastic TO-92 transistor package. The LM35D device is available in an 8-lead surface-mount small-outline package and a plastic TO-220 package.

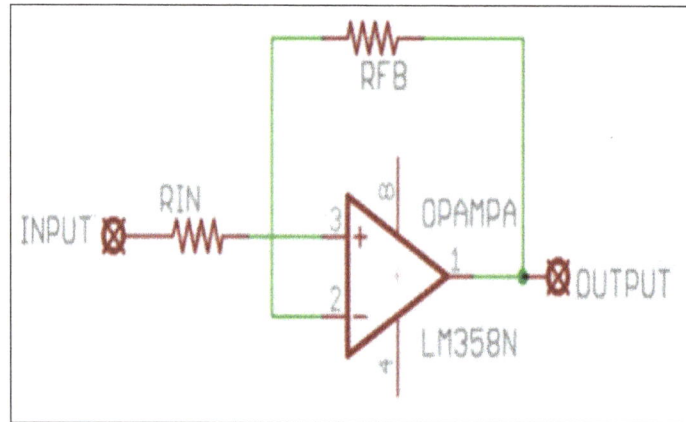

Circuit description of temperature sensor.

## LM35 Gain

Table: Gain for temperature sensor.

| Lm 35 | Ic | Av | %Error |
|-------|-------|-------|--------|
| 0.32V | 0.98V | 3.062 | -1.71 |
| 0.5V | 1.55V | 3.1 | -2.90 |
| 0.7V | 2.15V | 3.07 | -2 |
| 0.86V | 2.6V | 3.02 | -0.438 |
| 1V | 3.02 | 3.02 | -0.33 |

## Features

- Calibrated directly in celsius (Centigrade)

- Linear + 10-mV/ °C scale factor

- 0.5°C ensured accuracy (at 25 °C)

- Rated for full −55 °C to 150 °C range

- Suitable for remote applications

- Low-cost due to wafer-level trimming

## Humidity Sensor (Hy-Hs-220)

To measure humidity, amount of water molecules dissolved in the air of playhouse environments a smart humidity sensor module SY-HS-220 is opted for the system under design. The photograph of humidity sensor SY-HS-220 is shown in the figure. On close inspection of figure, it is found that the board consists of humidity sensor along with signal conditioning stages. The humidity sensor is of capacitive type, comprising on chip signal conditioner. However, it is mounted on the PCB which also consists of other stages employed to make sensor rather smarter. The PCB consists of CMOS timers to pulse the sensor to provide output voltage. Moreover, it also consists of oscillator, AC amplifier, frequency to voltage converter and precision rectifiers.

Humidity sensor

Incorporation of such stages on the board significantly helps to enhance the performance of the sensor. Moreover, it also helps to provide impediment to the noise. The humidity sensor used in this system is highly precise and reliable. It provides DC voltage depending upon humidity of the surrounding in RH%. This work with +5 Volt power supply and the typical current consumption is less than 3 mA. The operating humidity range is 30% RH to 90% RH. The standard DC output voltage provided at 250C is 1980 mV.

The accuracy is ± 5% RH at 250C. As shown in the figure, it provides three pins recognized as B, W and R. The pin labeled W provides the DC output voltage, where as the pin labeled B is ground. The VCC of +5V is applied at the pin R. The humidity dependent voltage is obtained and subjected for further processing. Signal Conditioner: As stated earlier, a smart humidity sensor SY-HS-220 provides the D.C. output voltage (mV) linearly.

## Features of Humidity Sensor

These modules convert the relative humidity to the output voltage.

Table: Features of humidity sensor.

| S. No | Specification | Range |
|-------|---------------|-------|
| 1 | Rated Voltage | DC 5.0V |
| 2 | Current Consumtion | <-3.0Ma |
| 3 | Operating Temperature Range | 0.60 °C |
| 4 | Operating Humidity Range | 30-90%RH |
| 5 | Storable Temperature Range | -30 °C~85 °C |
| 6 | Storable Humidity Range | Within 95%RH |
| 7 | Standard Output Range | DC 1.980Mv(at 25 °C,60%RH) |
| 8 | Accuracy | ±5%RH(at 25 °C,60%RH) |

## CO Sensor (MQ-7)

This is a simple-to-use Carbon Monoxide (CO) sensor suitable for sensing CO concentrations in the air. The MQ-7 can detect CO-gas concentrations anywhere from 20 to 2000ppm.

This sensor has a high sensitivity and fast response time. The sensor's output is an analog resistance. The drive circuit is very simple all you need to do is power the heater coil with 5V adds a load resistance and connects the output to an ADC.

CO sensor (MQ-7)

Sensitive material of MQ-7 gas sensor is $SnO_2$, which with lower conductivity in clean air. It make detection by method of cycle high and low temperature, and detect CO when low temperature (heated by 1.5V). The sensor's conductivity is higher along with the gas concentration rising. When high temperature (heated by 5.0V), it cleans the other gases adsorbed under low temperature. Please use simple electro circuit, Convert change of conductivity to correspond output signal of gas concentration. MQ-7 gas sensor has high sensity to Carbon Monoxide. The sensor could be used to detect different gases contains CO, it is with low cost and suitable for different application.

## Circuit Description

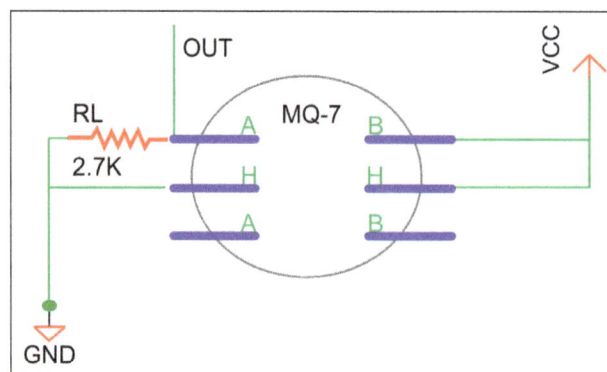

Circuit description of MQ-7

## Characteristics

- Good sensitivity to Combustible gas in wide range

- High sensitivity to Natural gas

- Long life and low cost

- Simple drive circuit

## Application

- Domestic gas leakage detector

- Industrial CO detector

- Portable gas detector

## Tabulation for Sensor Value

INPUT=5V (constant)

Table: Values of MQ7 sensor in various gases.

| S.no | Gas Range | Sensor output (PPM) |
|------|-----------|---------------------|
| 1 | No gas | 330 |
| 2 | Low gas | 450 |
| 3 | High gas | 900 |

## Carbon Dioxide Sensor (MQ-4)

Carbon dioxide is a non-toxic and non-flammable gas. However, exposure to elevated concentrations can include a risk to life. Whenever $CO_2$ gas or dry ice is used, produced, shipped or stored. $CO_2$ concentrations can rise to dangerously high levels. Because $CO_2$ is odourless and colourless leakages are impossible to detect, meaning proper sensors are needed to help ensure the safety.

Carbon dioxide sensor

Sensitive material of MQ-4 gas sensor is $SnO_2$, which with lower conductivity in clean air. It make detection by method of cycle high and low temperature and detect $CO_2$ when low temperature (heated by 1.5V).The sensor's conductivity is higher along with the gas concentration rising. When high temperature (heated by 5.0V), it detects Methane, Propane etc. Combustible gas and cleans the other gases absorbed under temperature.

## Temperature Considerations

The $CO_2$ gas sensor is somewhat sensitive to temperature changes. In most cases, variations in $CO_2$ readings due to temperature changes are small (<100 ppm on low range, <1000 ppm on high range).

With good experimental design, the variation in $CO_2$ readings due to a temperature change will be negligible compared to the overall change in $CO_2$ concentration. If an entire experiment is to be conducted at a constant temperature you could improve the accuracy of the readings by calibrating the sensor at that temperature. The sensor is designed to operate between 20 degree Celsius and 30 degree Celsius. It can be used outside of this temperature range.

However, the readings will be less accurate, even if you calibrate the sensor at the lower or higher temperature. This does not prohibit taking readings using incubation temperatures or outdoor readings at temperatures or outdoor readings at temperatures warmer or colder than the 20 to 30 degree Celsius range. Allow enough time for your $Co_2$ gas sensor to stabilize at the desired operating temperatures. The sensor is mainly based on the degree in Celsius.

## Circuit Description

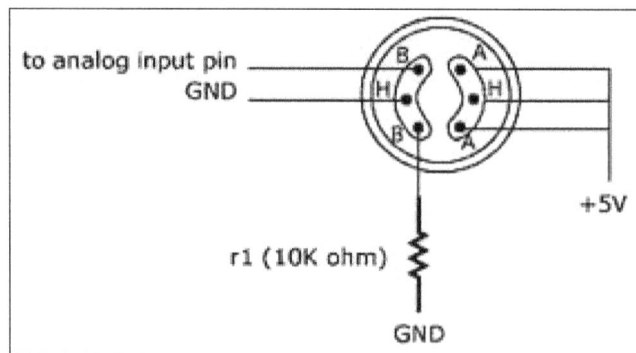

Circuit diagram of MQ-4

## Characteristics

- Good sensitivity to CO/Combustible gas
- High sensitivity to Methane, Propane and CO
- Long life and low cost
- Simple drive circuit

## Application

- Domestic gas leakage detector
- Industrial CO detector
- Industrial gas detector

# Air Quality Index

The AQI is an index for reporting daily air quality. It tells you how clean or polluted your air is, and what associated health effects might be a concern for you. The AQI focuses on health effects you may experience within a few hours or days after breathing polluted air. EPA calculates the AQI for five major air pollutants regulated by the Clean Air Act: ground-level ozone, particle pollution (also known as particulate matter), carbon monoxide, sulfur dioxide, and nitrogen dioxide. For each of these pollutants, EPA has established national air quality standards to protect public health. Ground-level ozone and airborne particles are the two pollutants that pose the greatest threat to human health in this country.

## How does the AQI Work?

Think of the AQI as a yardstick that runs from 0 to 500. The higher the AQI value, the greater the level of air pollution and the greater the health concern. For example, an AQI value of 50 represents good air quality with little potential to affect public health, while an AQI value over 300 represents hazardous air quality.

An AQI value of 100 generally corresponds to the national air quality standard for the pollutant, which is the level EPA has set to protect public health. AQI values below 100 are generally thought of as satisfactory. When AQI values are above 100, air quality is considered to be unhealthy-at first for certain sensitive groups of people, then for everyone as AQI values get higher.

## Understanding the AQI

The purpose of the AQI is to help you understand what local air quality means to your health. To make it easier to understand, the AQI is divided into six categories:

| Air Quality Index (AQI) Values | Levels of Health Concern | Colors |
|---|---|---|
| When the AQI is in this range: | Air quality conditions are: | As symbolized by this color: |
| 0 to 50 | Good | Green |
| 51 to 100 | Moderate | Yellow |
| 101 to 150 | Unhealthy for Sensitive Groups | Orange |
| 151 to 200 | Unhealthy | Red |
| 201 to 300 | Very Unhealthy | Purple |
| 301 to 500 | Hazardous | Maroon |

Each category corresponds to a different level of health concern. The six levels of health concern and what they mean are:

- "Good" AQI is 0 to 50: Air quality is considered satisfactory, and air pollution poses little or no risk.

- "Moderate" AQI is 51 to 100: Air quality is acceptable; however, for some pollutants there may be a moderate health concern for a very small number of people. For example, people who are unusually sensitive to ozone may experience respiratory symptoms.

- "Unhealthy for Sensitive Groups" AQI is 101 to 150: Although general public is not likely to be affected at this AQI range, people with lung disease, older adults and children are at a greater risk from exposure to ozone, whereas persons with heart and lung disease, older adults and children are at greater risk from the presence of particles in the air.

- "Unhealthy" AQI is 151 to 200: Everyone may begin to experience some adverse health effects, and members of the sensitive groups may experience more serious effects.

- "Very Unhealthy" AQI is 201 to 300: This would trigger a health alert signifying that everyone may experience more serious health effects.

- "Hazardous" AQI greater than 300: This would trigger a health warnings of emergency conditions. The entire population is more likely to be affected.

### AQI Colors

EPA has assigned a specific color to each AQI category to make it easier for people to understand quickly whether air pollution is reaching unhealthy levels in their communities. For example, the color orange means that conditions are "unhealthy for sensitive groups," while red means that conditions may be "unhealthy for everyone," and so on.

| Air Quality Index Levels of Health Concern | Numerical Value | Meaning |
|---|---|---|
| Good | 0 to 50 | Air quality is considered satisfactory, and air pollution poses little or no risk. |
| Moderate | 51 to 100 | Air quality is acceptable; however, for some pollutants there may be a moderate health concern for a very small number of people who are unusually sensitive to air pollution. |
| Unhealthy for Sensitive Groups | 101 to 150 | Members of sensitive groups may experience health effects. The general public is not likely to be affected. |
| Unhealthy | 151 to 200 | Everyone may begin to experience health effects; members of sensitive groups may experience more serious health effects. |
| Very Unhealthy | 201 to 300 | Health alert: everyone may experience more serious health effects. |
| Hazardous | 301 to 500 | Health warnings of emergency conditions. The entire population is more likely to be affected. |

Values above 500 are considered Beyond the AQI.

## Air Quality Health Index

The Air Quality Health Index or "AQHI" is a scale designed to help you understand what the air quality around you means to your health.

It is a health protection tool that is designed to help you make decisions to protect your health by limiting short-term exposure to air pollution and adjusting your activity levels during increased levels of air pollution. It also provides advice on how you can improve the quality of the air you breathe.

This index pays particular attention to people who are sensitive to air pollution and provides them with advice on how to protect their health during air quality levels associated with low, moderate, high and very high health risks.

The AQHI communicates four primary things:

1. Measures the air quality in relation to your health on a scale from 1 to 10. The higher the number, the greater the health risk associated with the air quality. When the amount of air pollution is very high, the number will be reported as 10+.

2. Assigns a category that describes the level of health risk associated with the index reading (e.g. Low, Moderate, High, or Very High Health Risk).

3. Provides health messages customized to each category for both the general population and the 'at risk' population.

4. Shows current hourly AQHI readings and maximum forecast values for today, tonight and tomorrow.

The AQHI is designed to give you this information along with some suggestions on how you might adjust your activity levels depending on your individual health risk from air pollution.

## How is the Air Quality Health Index Calculated?

The formula developed to calculate the Air Quality Health Index is based on research conducted by Health Canada using health and air quality data collected in major cities across Canada.

The Air Quality Health Index represents the relative risk of a mixture of common air pollutants which are known to harm human health. Three pollutants were chosen as indicators of the overall outdoor air mixture:

- Ground-level Ozone ($O_3$),

- Fine Particulate Matter ($PM_{2.5}$),

- Nitrogen Dioxide ($NO_2$).

## What is the scale for the Air Quality Health Index?

The Air Quality Health Index provides a number from 1 to 10+ to indicate the level of health risk associated with local air quality. Occasionally, when the amount of air pollution is abnormally high, the number may exceed 10.

The higher the number, the greater the health risk and our need to take precautions.

The index describes the level of health risk associated with this number as 'low', 'moderate', 'high' or 'very high', and suggests steps we can take to reduce our exposure.

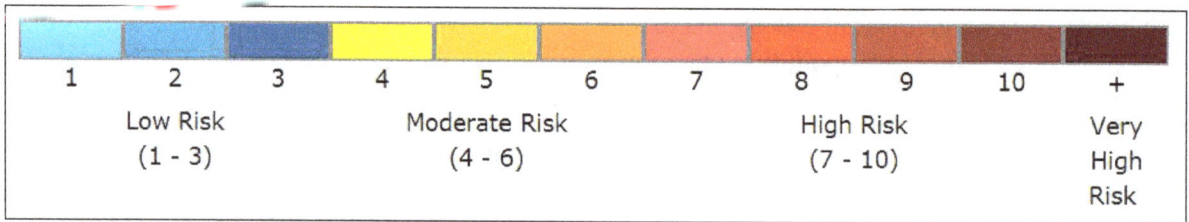

Air Quality Health Index Categories, Values and Associated Colours.

- 1-3 Low health risk

- 4-6 Moderate health risk

- 7-10 High health risk

- 10 + Very high health risk

## Air Quality Health Index Categories and Health Messages

The table below provides the health messages for each category of the Air Quality Health Index for the "at risk" population and the general population.

| Health Risk | Air Quality Health Index | Health Messages | |
|---|---|---|---|
| | | At Risk Population | General Population |
| Low | 1 - 3 | Enjoy your usual outdoor activities. | Ideal air quality for outdoor activities. |
| Moderate | 4 - 6 | Consider reducing or rescheduling strenuous activities outdoors if you are experiencing symptoms. | No need to modify your usual outdoor activities unless you experience symptoms such as coughing and throat irritation. |
| High | 7 - 10 | Reduce or reschedule strenuous activities outdoors. Children and the elderly should also take it easy. | Consider reducing or rescheduling strenuous activities outdoors if you experience symptoms such as coughing and throat irritation. |
| Very High | Above 10 | Avoid strenuous activities outdoors. Children and the elderly should also avoid outdoor physical exertion. | Reduce or reschedule strenuous activities outdoors, especially if you experience symptoms such as coughing and throat irritation. |

# 5

# Control and Prevention

Air pollution is controlled using various devices that prevent harmful gaseous and solid pollutants from entering the atmosphere. A few examples of such devices are dust collectors, scrubbers and thermal oxidizers. This chapter has been carefully written to provide an easy understanding of these types of air pollution control devices.

## Air Pollution Control Devices

Air pollution control devices are a series of devices that work to prevent a variety of different pollutants, both gaseous and solid, from entering the atmosphere primarily out of industrial smokestacks. These control devices can be separated into two broad categories-devices that control the amount of particulate matter escaping into the environment and devices that control acidic gas emissions. It is important to understand that the extraction methods for each specific type of pollutant can differ, so the only the major methods are discussed. Although complex, these devices have shown to be effective in the past with the overall levels of emissions for many pollutants dropping with the implementation of these control devices.

A chemical scrubber.

## Particulate Control

Specific machinery is used to remove particulate matter from flue gases. Much of this separation uses physical means of separation and not chemical separation techniques simply because particulate matter is large enough to be "caught" in this manner. Below are some of the basic ways that particulate matter can be extracted.

## Electrostatic Precipitators

An electrostatic precipitator is a type of filter that uses static electricity to remove soot and ash from exhaust fumes before they exit the smokestacks. Unburned particles of carbon in smoke are pulled out of the smoke by using static electricity in the precipitators, leaving clean, hot air to escape the smokestacks. It is vital to remove this unreacted carbon from the smoke, as it can damage buildings and harm human health - especially respiratory health.

## Cyclone Separators

A cyclone separator is a separation device that uses the principle of inertia to remove particulate matter from flue gases. In these separators, dirty flue gas enters a chamber containing a vortex, similar to a tornado. Because of the difference in inertia of gas particles and larger particulate matter, the gas particles move up the cylinder while larger particles hit the inside wall and drop down. This separates the particulate matter from the flue gas, leaving cleaned flue gas.

## Fabric Filters

Fabric filters are one fairly simple method that can be used to remove dust from flue gases. In some gases they can also remove acidic gases if they utilize basic compounds. This method simply uses some sort of fabric - generally felt is used as a woven cloth would allow dust to make its way through - is placed so that flue gasses must pass through it before exiting the smokestacks. When the gas passes through, dust particles are trapped in the cloth.

## Gas Control

More intense chemical methods of separation are generally required to separate polluting gases from the flue gas. However, this extraction is important as many acidic gases in flue gas contribute to acid rain. Below are some of the basic ways that gases can be extracted.

## Scrubbers

Scrubbers are a type of system that is used to remove harmful materials from industrial exhaust gases before they are released into the environment. These pollutants are generally gaseous, and when scrubbers are used to specifically remove SOx it is referred to as flue gas desulfurization. There are two main types of scrubbers, wet scrubbers and dry scrubbers. The main difference is in the type of material used to remove the gases. By removing acidic gases from the exhaust before it is released into the sky, scrubbers help prevent the formation of acid rain.

# Incineration

Incineration is used to convert VOC emissions into carbon dioxide and water through combustion. The incineration generally takes place in a specialized piece of equipment known as an afterburner, which is built to create the conditions necessary for complete combustion (such as sufficient burn time and a high temperature). Additionally, the incinerated gas must be mixed to ensure complete combustion.

# Carbon Capture

Carbon dioxide can theoretically also be captured and stored underground or in forests and oceans to prevent it from entering the atmosphere. Carbon capture and storage refers to the process of capturing this carbon dioxide and storing it below ground, pumping it into geologic layers. This process is rarely being used, but is talked about extensively as a way to limit greenhouse gas emissions leading to climate change.

# Dust Collector

A dust collector is a system used to enhance the quality of air released from industrial and commercial processes by collecting dust and other impurities from air or gas. Designed to handle high-volume dust loads, a dust collector system consists of a blower, dust filter, a filter-cleaning system, and a dust receptacle or dust removal system. It is distinguished from air purifiers, which use disposable filters to remove dust.

## Uses

Dust collectors are used in many processes to either recover valuable granular solid or powder from process streams, or to remove granular solid pollutants from exhaust gases prior to venting to the atmosphere. Dust collection is an online process for collecting any process-generated dust from the source point on a continuous basis. Dust collectors may be of single unit construction, or a collection of devices used to separate particulate matter from the process air. They are often used as an air pollution control device to maintain or improve air quality.

Mist collectors remove particulate matter in the form of fine liquid droplets from the air. They are often used for the collection of metal working fluids, and coolant or oil mists. Mist collectors are often used to improve or maintain the quality of air in the workplace environment.

Fume and smoke collectors are used to remove sub-micrometer-size particulates from the air. They effectively reduce or eliminate particulate matter and gas streams from many industrial processes such as welding, rubber and plastic processing, high speed machining with coolants, tempering, and quenching.

## Inertial Separators

Inertial separators separate dust from gas streams using a combination of forces, such as centrifugal, gravitational, and inertial. These forces move the dust to an area where the forces exerted by the gas stream are minimal. The separated dust is moved by gravity into a hopper, where it is temporarily stored.

The three primary types of inertial separators are:

- Settling chambers
- Baffle chambers
- Centrifugal collectors.

Neither settling chambers nor baffle chambers are commonly used in the minerals processing industry. However, their principles of operation are often incorporated into the design of more efficient dust collectors.

## Settling Chamber

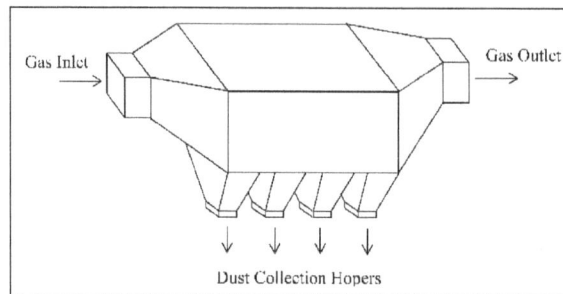

Settling Chamber.

A settling chamber consists of a large box installed in the ductwork. The increase of cross section area at the chamber reduces the speed of the dust-filled airstream and heavier particles settle out. Settling chambers are simple in design and can be manufactured from almost any material. However, they are seldom used as primary dust collectors because of their large space requirements and low efficiency. A practical use is as precleaners for more efficient collect. Advantages: 1) simple construction and low cost 2) maintenance free 3) collects particles without need of water. Disadvantages: 1) low efficiency 2) large space required.

## Baffle Chamber

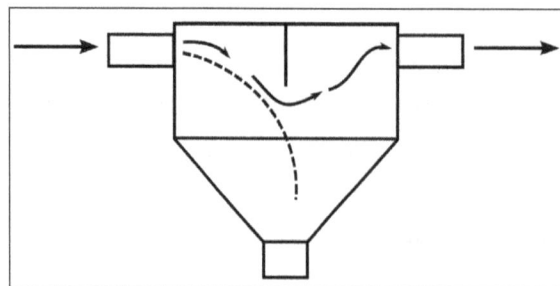

Baffle Chamber.

Baffle chambers use a fixed baffle plate that causes the conveying gas stream to make a sudden change of direction. Large-diameter particles do not follow the gas stream but continue into a dead air space and settle. Baffle chambers are used as precleaners.

## Centrifugal Collectors

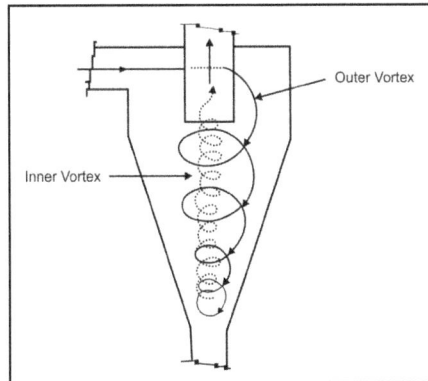

Cyclone.

Centrifugal collectors use cyclonic action to separate dust particles from the gas stream. In a typical cyclone, the dust gas stream enters at an angle and is spun rapidly. The centrifugal force created by the circular flow throws the dust particles toward the wall of the cyclone. After striking the wall, these particles fall into a hopper located underneath.

The most common types of centrifugal, or inertial, collectors in use today are:

## Single-cyclone Separators

Single-cyclone separators create a dual vortex to separate coarse from fine dust. The main vortex spirals downward and carries most of the coarser dust particles. The inner vortex, created near the bottom of the cyclone, spirals upward and carries finer dust particles.

## Multiple-cyclone Separators

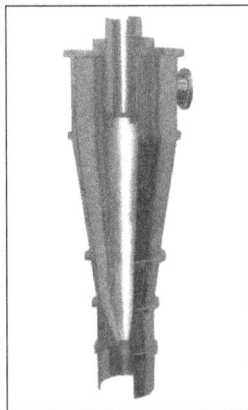

Multiclone.

Multiple-cyclone separators consist of a number of small-diameter cyclones, operating in parallel and having a common gas inlet and outlet, as shown in the figure, and operate on the same principle as single cyclone separators—creating an outer downward vortex and an ascending inner vortex.

Multiple-cyclone separators remove more dust than single cyclone separators because the individual cyclones have a greater length and smaller diameter. The longer length provides longer residence time while the smaller diameter creates greater centrifugal force. These two factors result in better separation of dust particulates. The pressure drop of multiple-cyclone separators collectors is higher than that of single-cyclone separators, requiring more energy to clean the same amount of air. A single-chamber cyclone separator of the same volume is more economical, but doesn't remove as much dust.

Cyclone separators are found in all types of power and industrial applications, including pulp and paper plants, cement plants, steel mills, petroleum coke plants, metallurgical plants, saw mills and other kinds of facilities that process dust.

## Secondary-air-flow Separators

This type of cyclone uses a secondary air flow, injected into the cyclone to accomplish several things. The secondary air flow increases the speed of the cyclonic action making the separator more efficient; it intercepts the particulate before it reaches the interior walls of the unit; and it forces the separated particulate toward the collection area. The secondary air flow protects the separator from particulate abrasion and allows the separator to be installed horizontally because gravity is not depended upon to move the separated particulate downward.

## Fabric Filters

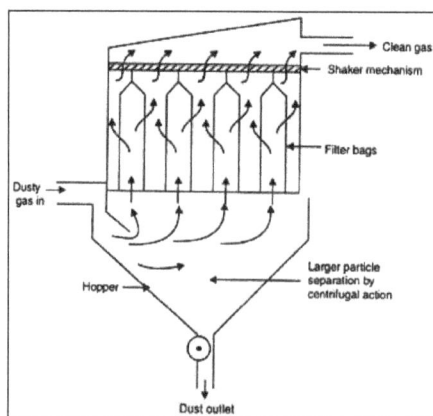

Baghouse.

Commonly known as baghouses, fabric collectors use filtration to separate dust particulates from dusty gases. They are one of the most efficient and cost-effective types of dust collectors available, and can achieve a collection efficiency of more than 99% for very fine particulates.

Dust-laden gases enter the baghouse and pass through fabric bags that act as filters. The bags can be of woven or felted cotton, synthetic, or glass-fiber material in either a tube or envelope shape.

## Pre-coating

To ensure the filter bags have a long usage life they are commonly coated with a filter enhancer (pre-coat). The use of chemically inert limestone (calcium carbonate) is most common as it maximises efficiency of dust collection (including fly ash) via formation of what is called a dustcake or coating on the surface of the filter media. This not only traps fine particulates but also provides protection for the bag itself from moisture, and oily or sticky particulates which can bind the filter media. Without a pre-coat the filter bag allows fine particulates to bleed through the bag filter system, especially during start-up, as the bag can only do part of the filtration leaving the finer parts to the filter enhancer dustcake.

## Parts

Fabric filters generally have the following parts:

1.  Clean plenum,

2.  Dusty plenum,

3.  Bag, cage and venturi assembly,

4.  Tubeplate,

5.  Rav/Screw,

6.  Compressed air header,

7.  Blow pipe,

8.  Housing and hopper.

## Types of Bag Cleaning

Baghouses are characterized by their cleaning method.

## Shaking

A rod connecting to the bag is powered by a motor. This provides motion to remove caked-on particles. The speed and motion of the shaking depends on the design of the bag and composition of the particulate matter. Generally shaking is horizontal. The top of the bag is closed and the bottom is open. When shaken, the dust collected on the inside of the bag is freed. No dirty gas flows through a bag while it is being cleaned. This redirection of air flow illustrates why baghouses must be compartmentalized.

### Reverse Air

Air flow gives the bag structure. Dirty air flows through the bag from the inside, allowing dust to collect on the interior surface. During cleaning, gas flow is restricted from a specific compartment. Without the flowing air, the bags relax. The cylindrical bag contains rings that prevent it from completely collapsing under the pressure of the air. A fan blows clean air in the reverse direction. The relaxation and reverse air flow cause the dust cake to crumble and release into the hopper. Upon the completion of the cleaning process, dirty air flow continues and the bag regains its shape.

### Pulse Jet

This type of baghouse cleaning (also known as pressure-jet cleaning) is the most common. A high pressure blast of air is used to remove dust from the bag. The blast enters the top of the bag tube, temporarily ceasing the flow of dirty air. The shock of air causes a wave of expansion to travel down the fabric. The flexing of the bag shatters and discharges the dust cake. The air burst is about 0.1 second and it takes about 0.5 seconds for the shock wave to travel down the length of the bag. Due to its rapid release, the blast of air does not interfere with contaminated gas flow. Therefore, pulse-jet bag-houses can operate continuously and are not usually compartmentalized. The blast of compressed air must be powerful enough to ensure that the shock wave will travel the entire length of the bag and fracture the dust cake. The efficiency of the cleaning system allows the unit to have a much higher gas to cloth ratio (or volumetric throughput of gas per unit area of filter) than shaking and reverse air bag filters. This kind of filter thus requires a smaller area to admit the same volume of air.

## Sonic

The least common type of cleaning method is sonic. Shaking is achieved by sonic vibration. A sound generator produces a low frequency sound that causes the bags to vibrate. Sonic cleaning is commonly combined with another method of cleaning to ensure thorough cleaning.

## Cartridge Collectors

Cartridge collectors use perforated metal cartridges that contain a pleated, nonwoven filtering media, as opposed to woven or felt bags used in baghouses. The pleated design allows for a greater total filtering surface area than in a conventional bag of the same diameter, The greater filtering area results in a reduced air to media ratio, pressure drop, and overall collector size.

Cartridge collectors are available in single use or continuous duty designs. In single-use collec-tors, the dirty cartridges are changed and collected dirt is removed while the collector is off. In the continuous duty design, the cartridges are cleaned by the conventional pulse-jet cleaning system.

## Selecting a Dust Collector

Dust collectors vary widely in design, operation, effectiveness, space requirements, construction, and capital, operating, and maintenance costs. Each type has advantages and disadvantages. How-ever, the selection of a dust collector should be based on the following general factors:

- Dust concentration and particle size – For minerals processing operations, the dust con-centration can range from 0.1 to 5.0 grains (0.32 g) of dust per cubic foot of air (0.23 to 11.44 grams per cubic meter), and the particle size can vary from 0.5 to 100 micrometres (μm) in diameter.

- Degree of dust collection required – The degree of dust collection required depends on its potential as a health hazard or public nuisance, the plant location, the allowable emission rate, the nature of the dust, its salvage value, and so forth. The selection of a collector should be based on the efficiency required and should consider the need for high-efficien-cy, high-cost equipment, such as electrostatic precipitators; high-efficiency, moderate-cost equipment, such as baghouses or wet scrubbers; or lower cost, primary units, such as dry centrifugal collectors.

- Characteristics of airstream – The characteristics of the airstream can have a significant impact on collector selection. For example, cotton fabric filters cannot be used where air temperatures exceed 180 °F (82 °C). Also, condensation of steam or water vapor can blind bags. Various chemicals can attack fabric or metal and cause corrosion in wet scrubbers.

- Characteristics of dust – Moderate to heavy concentrations of many dusts (such as dust from silica sand or metal ores) can be abrasive to dry centrifugal collectors. Hygroscopic material can blind bag collectors. Sticky material can adhere to collector elements and plug passages. Some particle sizes and shapes may rule out certain types of fabric collectors. The combustible nature of many fine materials rules out the use of electrostatic precipitators.

- Methods of disposal – Methods of dust removal and disposal vary with the material, plant process, volume, and type of collector used. Collectors can unload continuously or in batches. Dry materials can create secondary dust problems during unloading and disposal that do not occur with wet collectors. Disposal of wet slurry or sludge can be an additional material-handling problem; sewer or water pollution problems can result if wastewater is not treated properly.

## Fan and Motor

The fan and motor system supplies mechanical energy to move contaminated air from the dust-producing source to a dust collector.

## Types of Fans

There are two main kinds of industrial fans:

- Centrifugal fans,
- Axial-flow fans.

## Centrifugal Fans

Centrifugal fans consist of a wheel or a rotor mounted on a shaft that rotates in a scroll-shaped housing. Air enters at the eye of the rotor, makes a right-angle turn, and is forced through the blades of the rotor by centrifugal force into the scroll-shaped housing. The centrifugal force imparts static pressure to the air. The diverging shape of the scroll also converts a portion of the velocity pressure into static pressure.

There are three main types of centrifugal fans:

- Radial-blade fans - Radial-blade fans are used for heavy dust loads. Their straight, radial blades do not get clogged with material, and they withstand considerable abrasion. These fans have medium tip speeds and medium noise factors.

- Backward-blade fans - Backward-blade fans operate at higher tip speeds and thus are more efficient. Since material may build up on the blades, these fans should be used after a dust collector. Although they are noisier than radial-blade fans, backward-blade fans are commonly used for large-volume dust collection systems because of their higher efficiency.

- Forward-curved-blade fans - These fans have curved blades that are tipped in the direction of rotation. They have low space requirements, low tip speeds, and a low noise factor. They are usually used against low to moderate static pressures.

## Axial-flow Fans

Axial-flow fans are used in systems that have low resistance levels. These fans move the air parallel to the fan's axis of rotation. The screw-like action of the propellers moves the air in a straight-through parallel path, causing a helical flow pattern.

The three main kinds of axial fans are:

- Propeller fans - These fans are used to move large quantities of air against very low static pressures. They are usually used for general ventilation or dilution ventilation and are good in developing up to 0.5 in. wg (124.4 Pa).

- Tube-axial fans - Tube-axial fans are similar to propeller fans except they are mounted in a tube or cylinder. Therefore, they are more efficient than propeller fans and can develop up to 3 to 4 in. wg (743.3 to 995 Pa). They are best suited for moving air containing substances such as condensible fumes or pigments.

- Vane-axial fans - Vane-axial fans are similar to tube-axial fans except air-straightening vanes are installed on the suction or discharge side of the rotor. They are easily adapted to multistaging and can develop static pressures as high as 14 to 16 in. wg (3.483 to 3.98 kPa). They are normally used for clean air only.

## Fan Selection

When selecting a fan, the following points should be considered:

- Volume required.

- Fan static pressure.

- Type of material to be handled through the fan (For example, a radial-blade fan should be used with fibrous material or heavy dust loads, and nonsparking construction must be used with explosive or inflammable materials.).

- Type of drive arrangement, such as direct drive or belt drive.

- Space requirements.

- Noise levels.

- Operating temperature (For example, sleeve bearings are suitable to 250 °F/121.1 °C; ball bearings to 550 °F/287.8 °C).

- Sufficient size to handle the required volume and pressure with minimum horsepower.

- Need for special coatings or construction when operating in corrosive atmospheres.

- Ability of fan to accommodate small changes in total pressure while maintaining the necessary air volume.

- Need for an outlet damper to control airflow during cold starts (If necessary, the damper may be interlocked with the fan for a gradual start until steady-state conditions are reached).

## Fan Rating Tables

After the above information is collected, the actual selection of fan size and speed is usually made from a rating table published by the fan manufacturer. This table is known as a multirating table, and it shows the complete range of capacities for a particular size of fan.

Points to note:

- The multirating table shows the range of pressures and speeds possible within the limits of the fan's construction.

- A particular fan may be available in different construction classes (identified as class I through IV) relating to its capabilities and limits.

- For a given pressure, the highest mechanical efficiency is usually found in the middle third of the volume column.

- A fan operating at a given speed can have an infinite number of ratings (pressure and volume) along the length of its characteristic curve. However, when the fan is installed in a dust collection system, the point of rating can only be at the point at which the system resistance curve intersects the fan characteristic curve.

- In a given system, a fan at a fixed speed or at a fixed blade setting can have a single rating only. This rating can be changed only be changing the fan speed, blade setting, or the system resistance.

- For a given system, an increase in exhaust volume will result in increases in static and total pressures. For example, for a 20% increase in exhaust volume in a system with 5 in. pressure loss, the new pressure loss will be $5 \times (1.20)^2 = 7.2$ in.

- For rapid estimates of probable exhaust volumes available for a given motor size, the equation for brake horsepower, as illustrated, can be useful.

Fan installation Typical fan discharge conditions Fan ratings for volume and static pressure, as described in the multirating tables, are based on the tests conducted under ideal conditions. Often, field installation creates airflow problems that reduce the fan's air delivery. The following points should be considered when installing the fan:

- Avoid installation of elbows or bends at the fan discharge, which will lower fan performance by increasing the system's resistance.

- Avoid installing fittings that may cause non-uniform flow, such as an elbow, mitred elbow, or square duct.

- Check that the fan impeller is rotating in the proper direction-clockwise or counterclockwise.

- For belt-driven fans:
  - Check that the motor sheave and fan sheave are aligned properly.
  - Check for proper belt tension.

- Check the passages between inlets, impeller blades, and inside of housing for buildup of dirt, obstructions, or trapped foreign matter.

## Electric Motors

Electric motors are used to supply the necessary energy to drive the fan.

Integral-horsepower electric motors are normally three-phase, alternating-current motors. Fractional-horsepower electric motors are normally single-phase, alternating-current motors and are used when less than 1 hp (0.75 kW) is required. Since most dust collection systems require motors with more than 1 hp (0.75 kW), only integral-horsepower motors are discussed here.

The two most common types of integral-horsepower motors used in dust collection systems are:

- Squirrel-cage motors - These motors have a constant speed and are of a nonsynchronous, induction type.

- Wound-rotor motors - These motors are also known as slip-ring motors. They are general-purpose or continuous-rated motors and are chiefly used when an adjustable-speed motor is desired.

Squirrel-cage and wound-rotor motors are further classified according to the type of enclosure they use to protect their interior windings. These enclosures fall into two broad categories:

- Open,

- Totally enclosed.

Drip-proof and splash-proof motors are open motors. They provide varying degrees of protection; however, they should not be used where the air contains substances that might be harmful to the interior of the motor.

Totally enclosed motors are weather-protected with the windings enclosed. These enclosures prevent free exchange of air between the inside and the outside, but they are not airtight.

Totally enclosed, fan-cooled (TEFC) motors are another kind of totally enclosed motor. These motors are the most commonly used motors in dust collection systems. They have an integral-cooling fan outside the enclosure, but within the protective shield, that directs air over the enclosure.

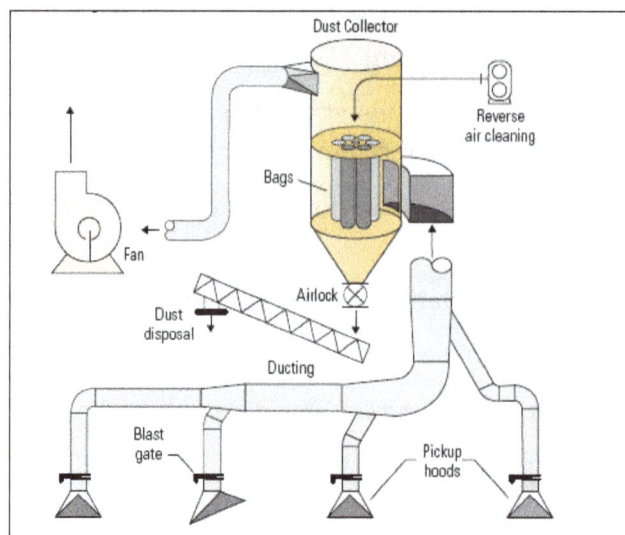

Dust collection system example.

Both open and totally enclosed motors are available in explosion-proof and dust-ignition-proof models to protect against explosion and fire in hazardous environments.

Motors are selected to provide sufficient power to operate fans over the full range of process conditions (temperature and flow rate).

## Configurations

Dust collectors can be configured into one of five common types:

- Ambient units - Ambient units are free-hanging systems for use when applications limit the use of source-capture arms or ductwork.

- Collection booths - Collector booths require no ductwork, and allow the worker greater freedom of movement. They are often portable.

- Downdraft tables - A downdraft table is a self-contained portable filtration system that removes harmful particulates and returns filtered air back into the facility with no external ventilation required.

- Source collector or Portable units - Portable units are for collecting dust, mist, fumes, or smoke at the source.

- Stationary units - An example of a stationary collector is a baghouse.

## Parameters involved in Specifying Dust Collectors

Important parameters in specifying dust collectors include airflow the velocity of the air stream created by the vacuum producer; system power, the power of the system motor, usually specified in horsepower; storage capacity for dust and particles, and minimum particle size filtered by the unit. Other considerations when choosing a dust collection system include the temperature, moisture content, and the possibility of combustion of the dust being collected.

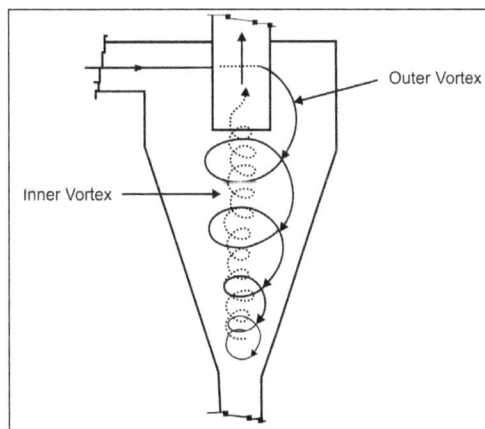

A cyclone separator is an apparatus for the separation, by centrifugal means, of fine particles suspended in air or gas.

Systems for fine removal may only contain a single filtration system (such as a filter bag or cartridge). However, most units utilize a primary and secondary separation/filtration system. In many cases the heat or moisture content of dust can negatively affect the filter media of a baghouse or cartridge dust collector. A cyclone separator or dryer may be placed before these units to reduce heat or moisture content before reaching the filters. Furthermore, some units may have third and fourth stage filtration. All separation and filtration systems used within the unit should be specified.

A baghouse is an air pollution abatement device used to trap particulate by filtering gas streams through large fabric bags. They are typically made of glass fibers or fabric.

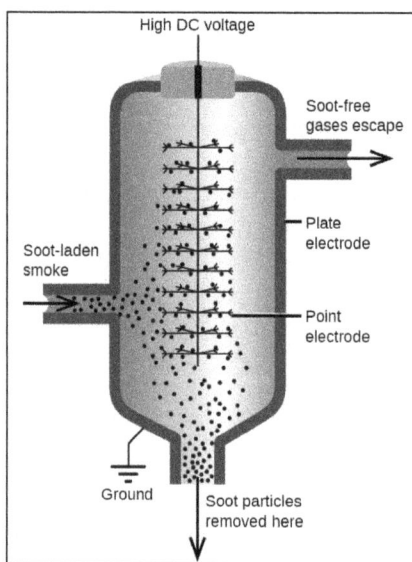

Electrostatic Precipitator

Electrostatic precipitators are a type of air cleaner, which charges particles of dust by passing dust-laden air through a strong (50-100 kV) electrostatic field. This causes the particles to be attracted to oppositely charged plates so that they can be removed from the air stream.

An impinger system is a device in which particles are removed by impacting the aerosol particles into a liquid. Modular media type units combine a variety of specific filter modules in one unit. These systems can provide solutions to many air contaminant problems. A typical system incorporates a series of disposable or cleanable pre-filters, a disposable vee-bag or cartridge filter. HEPA or carbon final filter modules can also be added. Various models are available, including free-hanging or ducted installations, vertical or horizontal mounting, and fixed or portable configurations. Filter cartridges are made out of a variety of synthetic fibers and are capable of collecting sub-micrometre particles without creating an excessive pressure drop in the system. Filter cartridges require periodic cleaning.

A wet scrubber, or venturi scrubber, is similar to a cyclone but it has an orifice unit that sprays water into the vortex in the cyclone section, collecting all of the dust in a slurry system. The water media can be recirculated and reused to continue to filter the air. Eventually the solids must be removed from the water stream and disposed of.

## Filter Cleaning Methods

- Online cleaning – Automatically timed filter cleaning which allows for continuous, uninterrupted dust collector operation for heavy dust operations.

- Offline cleaning – Filter cleaning accomplished during dust collector shut down. Practical whenever the dust loading in each dust collector cycle does not exceed the filter capacity. Allows for maximum effectiveness in dislodging and disposing of dust.

- On-demand cleaning – Filter cleaning initiated automatically when the filter is fully loaded, as determined by a specified drop in pressure across the media surface.

- Reverse-pulse/Reverse-jet cleaning – Filter cleaning method which delivers blasts of compressed air from the clean side of the filter to dislodge the accumulated dust cake.

- Impact/Rapper cleaning – Filter cleaning method in which high-velocity compressed air forced through a flexible tube results in an arbitrary rapping of the filter to dislodge the dust cake. Especially effective when the dust is extremely fine or sticky.

## Electrostatic Precipitator

An electrostatic precipitator (ESP) is a filtration device that removes fine particles, like dust and smoke, from a flowing gas using the force of an induced electrostatic charge minimally impeding the flow of gases through the unit.

In contrast to wet scrubbers which apply energy directly to the flowing fluid medium, an ESP applies energy only to the particulate matter being collected and therefore is very efficient in its consumption of energy (in the form of electricity).

## Invention of the Electrostatic Precipitator

The first use of corona discharge to remove particles from an aerosol was by Hohlfeld in 1824. However, it was not commercialized until almost a century later.

In 1907 Frederick Gardner Cottrell, a professor of chemistry at the University of California, Berkeley, applied for a patent on a device for charging particles and then collecting them through electrostatic attraction—the first electrostatic precipitator. Cottrell first applied the device to the collection of sulphuric acid mist and lead oxide fumes emitted from various acid-making and smelting activities. Wine-producing vineyards in northern California were being adversely affected by the lead emissions.

At the time of Cottrell's invention, the theoretical basis for operation was not understood. The operational theory was developed later in Germany, with the work of Walter Deutsch and the formation of the Lurgi company.

Cottrell used proceeds from his invention to fund scientific research through the creation of a foundation called Research Corporation in 1912, to which he assigned the patents. The intent of the organization was to bring inventions made by educators (such as Cottrell) into the commercial world for the benefit of society at large. The operation of Research Corporation is funded by royalties paid by commercial firms after commercialization occurs. Research Corporation has provided vital funding to many scientific projects: Goddard's rocketry experiments, Lawrence's cyclotron, production methods for vitamins A and $B_1$, among many others.

Research Corporation set territories for manufacturers of this technology, which included Western Precipitation (Los Angeles), Lodge-Cottrell (England), Lurgi Apparatebau-Gesellschaft (Germany), and Japanese Cottrell Corp. (Japan), as well as was a clearinghouse for any process improvements However, anti-trust concerts forced Research Corporation eliminate territory restrictions in 1946.

Electrophoresis is the term used for migration of gas-suspended charged particles in a direct-current electrostatic field. Traditional CRT television sets tend to accumulate dust on the screen because of this phenomenon (a CRT is a direct-current machine operating at about 15 kilovolts).

## Plate Precipitator

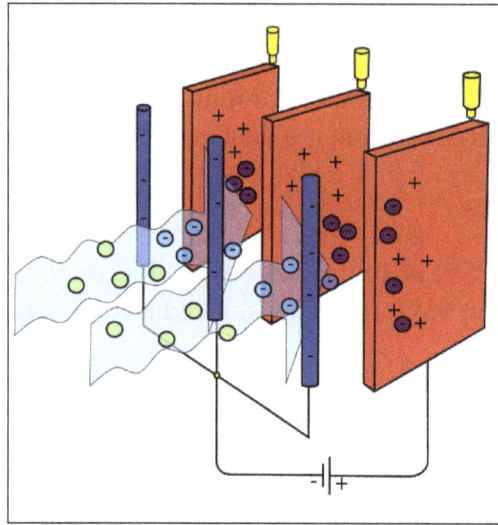

Conceptual diagram of an electrostatic precipitator.

The most basic precipitator contains a row of thin vertical wires, and followed by a stack of large flat metal plates oriented vertically, with the plates typically spaced about 1 cm to 18 cm apart, depending on the application. The air stream flows horizontally through the spaces between the wires, and then passes through the stack of plates.

A negative voltage of several thousand volts is applied between wire and plate. If the applied voltage is high enough, an electric corona discharge ionizes the air around the electrodes, which then ionizes the particles in the air stream.

The ionized particles, due to the electrostatic force, are diverted towards the grounded plates. Particles build up on the collection plates and are removed from the air stream.

A two-stage design (separate charging section ahead of collecting section) has the benefit of minimizing ozone production, which would adversely affect health of personnel working in enclosed spaces. For shipboard engine rooms where gearboxes generate an oil mist, two-stage ESP's are used to clean the air, improving the operating environment and preventing buildup of flammable oil fog accumulations. Collected oil is returned to the gear lubricating system.

## Collection Efficiency (R)

Precipitator performance is very sensitive to two particulate properties: 1) Electrical resistivity; and 2) Particle size distribution. These properties can be measured economically and accurately in the laboratory, using standard tests. Resistivity can be determined as a function of temperature in accordance with IEEE Standard 548. This test is conducted in an air environment containing a specified moisture concentration. The test is run as a function of ascending or descending temperature, or both. Data is acquired using an average ash layer electric field of 4 kV/cm. Since relatively low applied voltage is used and no sulfuric acid vapor is present in the test environment, the values obtained indicate the maximum ash resistivity.

In an ESP, where particle charging and discharging are key functions, resistivity is an important factor that significantly affects collection efficiency. While resistivity is an important phenomenon in the inter-electrode region where most particle charging takes place, it has a particularly important effect on the dust layer at the collection electrode where discharging occurs. Particles that exhibit high resistivity are difficult to charge. But once charged, they do not readily give up their acquired charge on arrival at the collection electrode. On the other hand, particles with low resistivity easily become charged and readily release their charge to the grounded collection plate. Both extremes in resistivity impede the efficient functioning of ESPs. ESPs work best under normal resistivity conditions.

Resistivity, which is a characteristic of particles in an electric field, is a measure of a particle's resistance to transferring charge (both accepting and giving up charges). Resistivity is a function of a particle's chemical composition as well as flue gas operating conditions such as temperature and moisture. Particles can have high, moderate (normal), or low resistivity.

Bulk resistivity is defined using a more general version of Ohm's Law, as given in equation below:

$$\vec{E} = \rho \vec{j}$$

where:

$E$ is the Electric field strength (V/cm).

$j$ is the Current density (A/cm$^2$).

$\rho$ is the Resistivity (Ohm-cm).

A better way of displaying this would be to solve for resistivity as a function of applied voltage and current, as given in equation below:

$$\rho = \frac{AV}{Il}$$

where:

$\rho$ = Resistivity (Ohm-cm)

V = The applied DC potential, (Volts).

I = The measured current, (Amperes).

l = The ash layer thickness, (cm).

A = The current measuring electrode face area, (cm$^2$).

Resistivity is the electrical resistance of a dust sample 1.0 cm$^2$ in cross-sectional area, 1.0 cm thick, and is recorded in units of ohm-cm. The table below, gives value ranges for low, normal, and high resistivity.

| Resistivity | Range of Measurement |
|---|---|
| Low | Between $10^4$ and $10^7$ ohm-cm |
| Normal | Between $10^7$ and $2 \times 10^{10}$ ohm-cm |
| High | Above $2 \times 10^{10}$ ohm-cm |

## Dust Layer Resistance

Resistance affects electrical conditions in the dust layer by a potential electric field (voltage drop) being formed across the layer as negatively charged particles arrive at its surface and leak their electrical charges to the collection plate. At the metal surface of the electrically grounded collection plate, the voltage is zero, whereas at the outer surface of the dust layer, where new particles and ions are arriving, the electrostatic voltage caused by the gas ions can be quite high. The strength of this electric field depends on the resistance and thickness of the dust layer.

In high-resistance dust layers, the dust is not sufficiently conductive, so electrical charges have difficulty moving through the dust layer. Consequently, electrical charges accumulate on and beneath the dust layer surface, creating a strong electric field.

Voltages can be greater than 10,000 volts. Dust particles with high resistance are held too strongly to the plate, making them difficult to remove and causing rapping problems.

In low resistance dust layers, the corona current is readily passed to the grounded collection electrode. Therefore, a relatively weak electric field, of several thousand volts, is maintained across the dust layer. Collected dust particles with low resistance do not adhere strongly enough to the collection plate. They are easily dislodged and become retained in the gas stream.

The electrical conductivity of a bulk layer of particles depends on both surface and volume factors. Volume conduction, or the motions of electrical charges through the interiors of particles, depends mainly on the composition and temperature of the particles. In the higher temperature regions, above 500 °F (260 °C), volume conduction controls the conduction mechanism. Volume conduction also involves ancillary factors, such as compression of the particle layer, particle size and shape, and surface properties.

Volume conduction is represented in the figures as a straight-line at temperatures above 500 °F (260 °C). At temperatures below about 450 °F (230 °C), electrical charges begin to flow across surface moisture and chemical films adsorbed onto the particles. Surface conduction begins to lower the resistivity values and bend the curve downward at temperatures below 500 °F (260 °C).

These films usually differ both physically and chemically from the interiors of the particles owing to adsorption phenomena. Theoretical calculations indicate that moisture films only a few molecules thick are adequate to provide the desired surface conductivity. Surface conduction on particles is closely related to surface-leakage currents occurring on electrical insulators, which have been extensively studied. An interesting practical application of surface-leakage is the determination of dew point by measurement of the current between adjacent electrodes mounted on a glass surface.

A sharp rise in current signals the formation of a moisture film on the glass. This method has been used effectively for determining the marked rise in dew point, which occurs when small amounts of sulfuric acid vapor are added to an atmosphere (commercial Dewpoint Meters are available on the market).

The following discussion of normal, high, and low resistance applies to ESPs operated in a dry state; resistance is not a problem in the operation of wet ESPs because of the moisture concentration in the ESP.

## Normal Resistivity

As stated above, ESPs work best under normal resistivity conditions. Particles with normal resistivity do not rapidly lose their charge on arrival at the collection electrode. These particles slowly leak their charge to grounded plates and are retained on the collection plates by intermolecular adhesive and cohesive forces. This allows a particulate layer to be built up and then dislodged from the plates by rapping. Within the range of normal dust resistivity (between $10^7$ and $2 \times 10^{10}$ ohm-cm), fly ash is collected more easily than dust having either low or high resistivity.

## High Resistivity

If the voltage drop across the dust layer becomes too high, several adverse effects can occur. First, the high voltage drop reduces the voltage difference between the discharge electrode and collection electrode, and thereby reduces the electrostatic field strength used to drive the gas ion-charged particles over to the collected dust layer. As the dust layer builds up, and the electrical charges accumulate on the surface of the dust layer, the voltage difference between the discharge and collection electrodes decreases. The migration velocities of small particles are especially affected by the reduced electric field strength.

Another problem that occurs with high resistivity dust layers is called back corona. This occurs when the potential drop across the dust layer is so great that corona discharges begin to appear in the gas that is trapped within the dust layer. The dust layer breaks down electrically, producing small holes or craters from which back corona discharges occur. Positive gas ions are generated within the dust layer and are accelerated toward the "negatively charged" discharge electrode. The positive ions reduce some of the negative charges on the dust layer and neutralize some of the negative ions on the "charged particles" heading toward the collection electrode. Disruptions of the normal corona process greatly reduce the ESP's collection efficiency, which in severe cases, may fall below 50%. When back corona is present, the dust particles build up on the electrodes forming a layer of insulation. Often this can not be repaired without bringing the unit offline.

The third, and generally most common problem with high resistivity dust is increased electrical sparking. When the sparking rate exceeds the "set spark rate limit," the automatic controllers limit the operating voltage of the field. This causes reduced particle charging and reduced migration velocities toward the collection electrode. High resistivity can generally be reduced by doing the following:

- Adjusting the temperature.

- Increasing moisture content.

- Adding conditioning agents to the gas stream.

- Increasing the collection surface area.

- Using hot-side precipitators (occasionally and with foreknowledge of sodium depletion).

Thin dust layers and high-resistivity dust especially favor the formation of back corona craters. Severe back corona has been observed with dust layers as thin as 0.1 mm, but a dust layer just over one particle thick can reduce the sparking voltage by 50%. The most marked effects of back corona on the current-voltage characteristics are:

- Reduction of the spark over voltage by as much as 50% or more.

- Current jumps or discontinuities caused by the formation of stable back-corona craters.

- Large increase in maximum corona current, which just below spark over corona gap may be several times the normal current.

The figure below and to the left shows the variation in resistivity with changing gas temperature for six different industrial dusts along with three coal-fired fly ashes. The figure on the right illustrates resistivity values measured for various chemical compounds that were prepared in the laboratory.

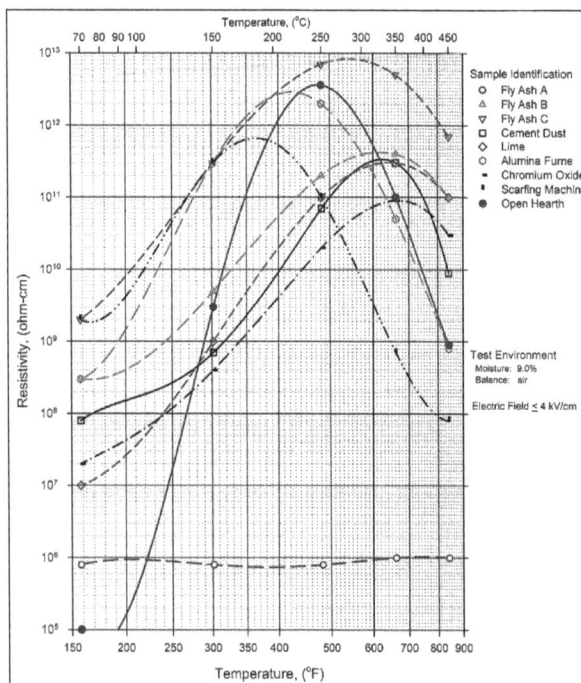

Resistivity values of representative dusts
and fumes from industrial plants.

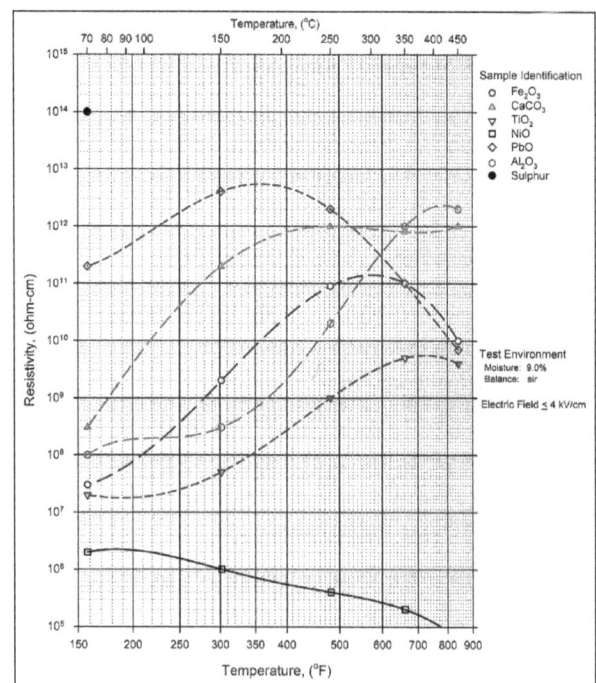

Resistivity values of various chemicals and
reagents as a function of temperature.

Results for Fly Ash A were acquired in the ascending temperature mode. These data are typical for a moderate to high combustibles content ash. Data for Fly Ash B are from the same sample, acquired during the descending temperature mode.

The differences between the ascending and descending temperature modes are due to the presence of unburned combustibles in the sample. Between the two test modes, the samples are equilibrated in dry air for 14 hours (overnight) at 850 °F (450 °C). This overnight annealing process typically removes between 60% and 90% of any unburned combustibles present in the samples. Exactly how carbon works as a charge carrier is not fully understood, but it is known to significantly reduce the resistivity of a dust.

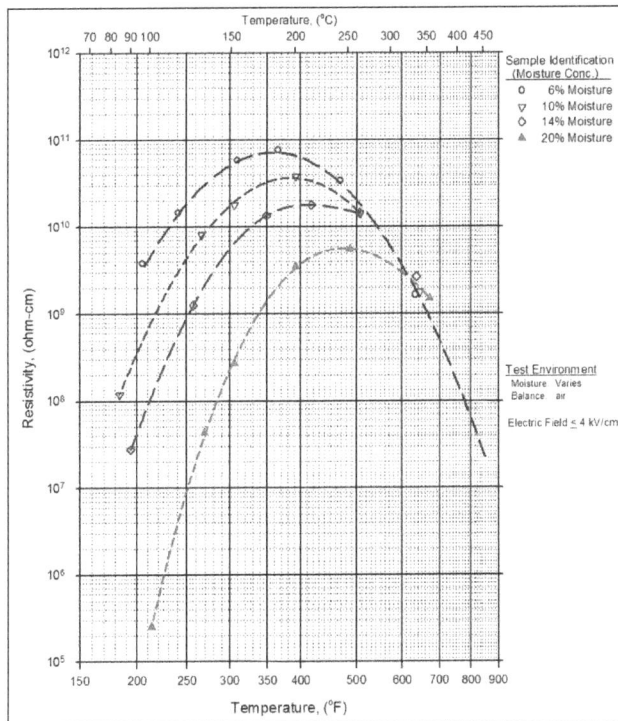

Resistivity measured as a function of temperature in
varying moisture concentrations (humidity).

Carbon can act, at first, like a high resistivity dust in the precipitator. Higher voltages can be required in order for corona generation to begin. These higher voltages can be problematic for the TR-Set controls. The problem lies in onset of corona causing large amounts of current to surge through the (low resistivity) dust layer. The controls sense this surge as a spark. As precipitators are operated in spark-limiting mode, power is terminated and the corona generation cycle re-initiates. Thus, lower power (current) readings are noted with relatively high voltage readings.

The same thing is believed to occur in laboratory measurements. Parallel plate geometry is used in laboratory measurements without corona generation. A stainless steel cup holds the sample. Another stainless steel electrode weight sits on top of the sample (direct contact with the dust layer). As voltage is increased from small amounts (e.g. 20 V), no current is measured. Then, a threshold voltage level is reached. At this level, current surges through the sample so much so that the voltage supply unit can trip off. After removal of the unburned combustibles during the above-mentioned annealing procedure, the descending temperature mode curve shows the typical inverted "V" shape one might expect.

## Low Resistivity

Particles that have low resistivity are difficult to collect because they are easily charged (very conductive) and rapidly lose their charge on arrival at the collection electrode. The particles take on the charge of the collection electrode, bounce off the plates, and become re-entrained in the gas stream. Thus, attractive and repulsive electrical forces that are normally at work at normal and higher resistivities are lacking, and the binding forces to the plate are considerably lessened. Examples of low-resistivity dusts are unburned carbon in fly ash and carbon black.

If these conductive particles are coarse, they can be removed upstream of the precipitator by using a device such as a cyclone mechanical collector.

The addition of liquid ammonia ($NH_3$) into the gas stream as a conditioning agent has found wide use in recent years. It is theorized that ammonia reacts with $H_2SO_4$ contained in the flue gas to form an ammonium sulfate compound that increases the cohesivity of the dust. This additional cohesivity makes up for the loss of electrical attraction forces.

The table below summarizes the characteristics associated with low, normal and high resistivity dusts.

The moisture content of the flue gas stream also affects particle resistivity. Increasing the moisture content of the gas stream by spraying water or injecting steam into the duct work preceding the ESP lowers the resistivity. In both temperature adjustment and moisture conditioning, one must maintain gas conditions above the dew point to prevent corrosion problems in the ESP or downstream equipment. The figure to the right shows the effect of temperature and moisture on the resistivity of a cement dust. As the percentage of moisture in the gas stream increases from 6 to 20%, the resistivity of the dust dramatically decreases. Also, raising or lowering the temperature can decrease cement dust resistivity for all the moisture percentages represented.

The presence of $SO_3$ in the gas stream has been shown to favor the electrostatic precipitation process when problems with high resistivity occur. Most of the sulfur content in the coal burned for combustion sources converts to $SO_2$. However, approximately 1% of the sulfur converts to $SO_3$. The amount of $SO_3$ in the flue gas normally increases with increasing sulfur content of the coal. The resistivity of the particles decreases as the sulfur content of the coal increases.

| Resistivity | Range of Measurement | Precipitator Characteristics |
| --- | --- | --- |
| Low | Between $10^4$ and $10^7$ ohm-cm | 1. Normal operating voltage and current levels unless dust layer is thick enough to reduce plate clearances and cause higher current levels.<br>2. Reduced electrical force component retaining collected dust, vulnerable to high reentrainment losses.<br>3. Negligible voltage drop across dust layer.<br>4. Reduced collection performance due to (2) |
| Normal | Between $10^7$ and $2 \times 10^{10}$ ohm-cm | 1. Normal operating voltage and current levels.<br>2. Negligible voltage drop across dust layer.<br>3. Sufficient electrical force component retaining collected dust.<br>4. High collection performance due to (1), (2) and (3) |

| Marginal to High | Between 2 x $10^{10}$ and $10^{12}$ ohm-cm | 1. Reduced operating voltage and current levels with high spark rates.<br>2. Significant voltage loss across dust layer.<br>3. Moderate electrical force component retaining collected dust.<br>4. Reduced collection performance due to (1) and (2) |
|---|---|---|
| High | Above $10^{12}$ ohm-cm | 1. Reduced operating voltage levels; high operating current levels if power supply controller is not operating properly.<br>2. Very significant voltage loss across dust layer.<br>3. High electrical force component retaining collected dust.<br>4. Seriously reduced collection performance due to (1), (2) and probably back corona. |

Other conditioning agents, such as sulfuric acid, ammonia, sodium chloride, and soda ash (sometimes as raw trona), have also been used to reduce particle resistivity. Therefore, the chemical composition of the flue gas stream is important with regard to the resistivity of the particles to be collected in the ESP. The table below lists various conditioning agents and their mechanisms of operation.

| Conditioning Agent | Mechanisms of Action |
|---|---|
| Sulfur Trioxide and Sulfuric Acid | 1. Condensation and adsorption on fly ash surfaces.<br>2. May also increase cohesiveness of fly ash.<br>3. Reduces resistivity |
| Ammonia | Mechanism is not clear, various ones proposed:<br>1. Modifies resistivity.<br>2. Increases ash cohesiveness.<br>3. Enhances space charge effect. |
| Ammonium Sulfate | Little is known about the mechanism; claims are made for the following:<br>1. Modifies resistivity (depends upon injection temperature).<br>2. Increases ash cohesiveness.<br>3. Enhances space charge effect.<br>4. Experimental data lacking to substantiate which of these is predominant. |
| Triethylamine | Particle agglomeration claimed; no supporting data. |
| Sodium Compounds | 1. Natural conditioner if added with coal.<br>2. Resistivity modifier if injected into gas stream. |
| Compounds of Transition Metals | Postulated that they catalyze oxidation of $SO_2$ to $SO_3$; no definitive tests with fly ash to verify this postulation. |
| Potassium Sulfate and Sodium Chloride | In cement and lime kiln ESPs:<br>1. Resistivity modifiers in the gas stream.<br>2. NaCl - natural conditioner when mixed with coal. |

If injection of ammonium sulfate occurs at a temperature greater than about 600 °F (320 °C), dissociation into ammonia and sulfur trioxide results. Depending on the ash, $SO_2$ may preferentially interact with fly ash as $SO_3$ conditioning. The remainder recombines with ammonia to add to the space charge as well as increase cohesiveness of the ash.

More recently, it has been recognized that a major reason for loss of efficiency of the electrostatic precipitator is due to particle buildup on the charging wires in addition to the collection plates. This is easily remedied by making sure that the wires themselves are cleaned at the same time that the collecting plates are cleaned.

Sulfuric acid vapor ($SO_3$) enhances the effects of water vapor on surface conduction. It is physically adsorbed within the layer of moisture on the particle surfaces. The effects of relatively small amounts of acid vapor can be seen in the figure below.

The inherent resistivity of the sample at 300 °F (150 °C) is $5 \times 10^{12}$ ohm-cm. An equilibrium concentration of just 1.9 ppm sulfuric acid vapor lowers that value to about $7 \times 10^9$ ohm-cm.

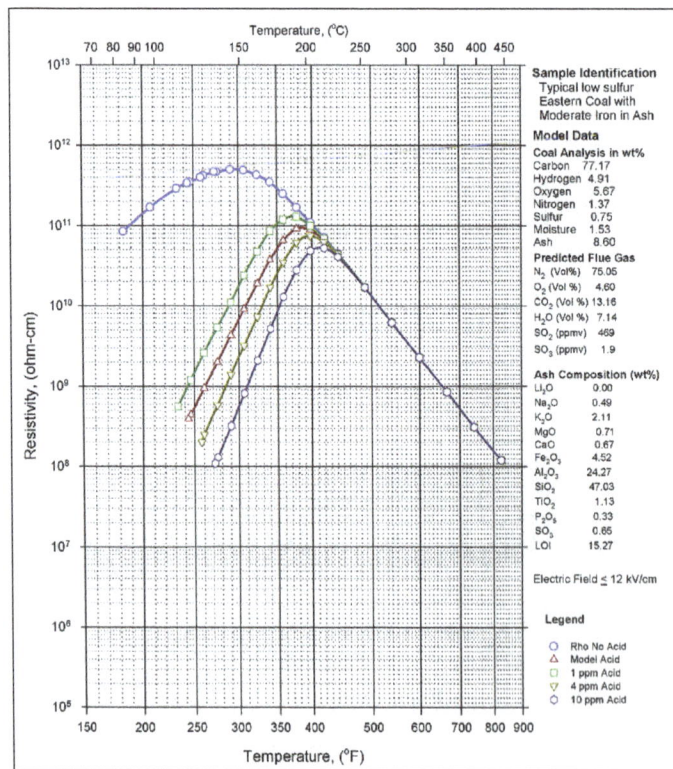

Resistivity modeled as a function of environmental conditions - especially sulfuric acid vapor.

## Modern Industrial Electrostatic Precipitators

ESPs continue to be excellent devices for control of many industrial particulate emissions, including smoke from electricity-generating utilities (coal and oil fired), salt cake collection from black liquor boilers in pulp mills, and catalyst collection from fluidized bed catalytic cracker units in oil refineries to name a few. These devices treat gas volumes from several hundred thousand ACFM to 2.5 million ACFM (1,180 m³/s) in the largest coal-fired boiler applications. For a coal-fired boiler the collection is usually performed downstream of the air preheater at about 160 °C (320 °F) which provides optimal resistivity of the coal-ash particles. For some difficult applications with low-sulfur fuel hot-end units have been built operating above 370 °C (698 °F).

A smokestack at coal-fired hazelwood power station in victoria, australia emits brown smoke when its esp is shut down.

The original parallel plate–weighted wire design has evolved as more efficient (and robust) discharge electrode designs were developed, today focusing on rigid (pipe-frame) discharge electrodes to which many sharpened spikes are attached (barbed wire), maximizing corona production. Transformer-rectifier systems apply voltages of 50–100 kV at relatively high current densities. Modern controls, such as an automatic voltage control, minimize electric sparking and prevent arcing (sparks are quenched within 1/2 cycle of the TR set), avoiding damage to the components. Automatic plate-rapping systems and hopper-evacuation systems remove the collected particulate matter while on line, theoretically allowing ESPs to stay in continuous operation for years at a time.

## Electrostatic Sampling for Bioaerosols

Electrostatic precipitators can be used to sample biological airborne particles or aerosol for analysis. Sampling for bioaerosols requires precipitator designs optimised with a liquid counter electrode, which can be used to sample biological particles, e.g. viruses, directly into a small liquid volume to reduce unnecessary sample dilution.

## Wet Electrostatic Precipitator

A wet electrostatic precipitator (WESP or wet ESP) operates with water vapor saturated air streams (100% relative humidity). WESPs are commonly used to remove liquid droplets such as sulfuric acid mist from industrial process gas streams. The WESP is also commonly used where the gases are high in moisture content, contain combustible particulate, or have particles that are sticky in nature.

## Consumer-oriented Electrostatic Air Cleaners

Plate precipitators are commonly marketed to the public as air purifier devices or as a permanent replacement for furnace filters, but all have the undesirable attribute of being somewhat messy to clean. A negative side-effect of electrostatic precipitation devices is the potential production of toxic ozone and $NO_x$. However, electrostatic precipitators offer benefits over other air purifications technologies, such as HEPA filtration, which require expensive filters and can become "production sinks" for many harmful forms of bacteria.

A portable electrostatic air cleaner
marketed to consumers.

Portable electrostatic air cleaner with cover
removed, showing collector plates.

With electrostatic precipitators, if the collection plates are allowed to accumulate large amounts of particulate matter, the particles can sometimes bond so tightly to the metal plates that vigorous washing and scrubbing may be required to completely clean the collection plates. The close spacing of the plates can make thorough cleaning difficult, and the stack of plates often cannot be easily disassembled for cleaning. One solution, suggested by several manufacturers, is to wash the collector plates in a dishwasher.

Some consumer precipitation filters are sold with special soak-off cleaners, where the entire plate array is removed from the precipitator and soaked in a large container overnight, to help loosen the tightly bonded particulates.

# Scrubbers

A scrubber or scrubber system is a system that is used to remove harmful materials from industrial exhaust gases before they are released into the environment. There are two main ways to scrub pollutants out of exhaust, and they are:

- Wet Scrubbing: The removal of harmful components of exhausted flue gases by spraying a liquid substance through the gas.

- Dry Scrubbing: The removal of harmful components of exhausted flue gases by introducing a solid substance to the gas - generally in powdered form.

Both of these methods work similarly and perform the same process of removing pollutants. The main difference is the materials they use to filter the gases. By removing acidic gases from the exhaust before it is released into the sky, scrubbers help prevent the formation of acid rain.

A chemical scrubber.

## Use

Scrubbing, sometimes referred to as flue gas desulfurization is the most effect sulfur-removal technique that is in widespread use. Removing the sulfur oxides is fairly simple, the flue gases pass through a spray of water in a wet scrubber that contains a variety of chemicals. Generally speaking, the main chemical is calcium carbonate. If a dry scrubber is used, the flue gas comes into contact with pulverized limestone - which is mainly calcium carbonate. The chemical reaction between the calcium carbonate and the sulfur dioxide yield calcium sulfite. This calcium sulfite either falls out of the gas stream or is removed with other particulates.

Scrubbers are very effective, removing about 98% of sulfur from flue gases, but they are very expensive to maintain and install. Additionally, it is energy intensive as the flue gas must be reheated after coming into contact with water vapour in the wet scrubber for the gas to be buoyant enough to exit through the smokestacks.

## Environmental Impacts

The use of scrubbers to clean flue gases before they leave the smokestacks has a drastic, beneficial impact on the environment. By collecting particulate matter and acidic gases, the amount of different pollutants that can exit the plant and be introduced into the environment is dramatically reduced. This increases air quality and lowers the health risks for people who could come into contact with the different pollutants.

Although there are many positive side-effects of using scrubbers, there are still waste products from the scrubbing process whether wet or dry scrubbing is used. These by-products must be disposed of safely since they can rarely be reused because of their chemical content. This is one reason that dry scrubbing has become more common, as the sheer volume of the waste products is less significant than the waste from a wet scrubbing operation.

# Gas Scrubber

Gas scrubber are cleanings installations in which the gas flow is brought in intensive contact with a fluid which as aim to remove gaseous components from the gas to the fluid. Gas scrubber can be applied as emission control technique at various gaseous emissions. Gas scrubbing is also called absorption.

At gas scrubbers there therefore talk of a transition of components of the gas phase to the Liquid phase. The degree in which is possible to convert gaseous components to the fluid phase depends on solubility on these components in the fluid. The balance concentration at the mist stage which belongs to a certain concentration at the fluid stage depends on the temperature, a higher temperature of the fluid phase results in a higher balance concentration at the mist stage. Lowering of the temperature has therefore also a favorable impact on the output. By adding chemicals to the carrier fluid in which absorbed components are converted, the load can increase. Adding of chemicals that can react with the absorbed gases has a favorable impact on the absorption output has.

Beside water (wet scrubbers) also organic fluids are used as an absorption means. In many cases chemicals or micro-organism are added to the carrier fluid to put or neutralize the gases which are solved in the fluid (conditioned scrubber). By this conversion the concentration in water becomes lower and more of the gasses can solve (according to the law of Henry). In practice gas scrubber exist from three components: an absorption section for substance exchange, demister and a recirculation tank with pump.

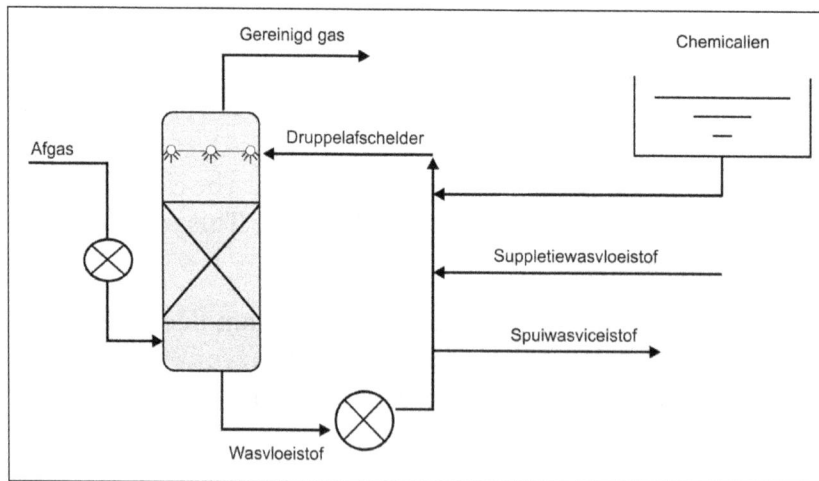

Gaswassing.

The liquid gas ratio L/G of the gas scrubber is the proportion between the flow of the carrier fluid and the flow of the gas flow. Concerning the dimensioning and for the appraisal of the functioning of the gas scrubber it is important know how much fluid by m³ gas is necessary to reach the desired remaining emission. L/G proportion is not only stipulated by required the remaining emission, but also in the concentration of the removable components in the gas flow and in - and outgoing liquid flows. L/G proportion in a concrete situation depends therefore on the chosen carrier system, the properties of the to purify gas, the carrier fluid and the removable component and the demands which are made to the remaining emission. Gas scrubber can be classified to the flow direction of the gas regarding to the fluid. Thereby it is distinguished in counter-, crossflowscrubbers or flowscrubber.

Gas scrubber can be also classified to the implementation of the transmission section, as it happens with or without deduction. The built-in can be a dumped or a structured packing is or a construction with plates or a rotating disk.

Applicability of the different types of scrubbers is especially stipulated by properties of the gas to purify.

The advantages of gas scrubbing are:

- Broad application range.
- Very high disposal output.
- Compact installation and simply in maintenance.
- Relatively simple technology.
- Can serve also as refrigeration for warm gas flows (quencher).

The disadvantages of the gas scrubbing are:

- Effluent must be treated.
- Water - and reactant consumption.
- When dust becomes simultaneous catched from, is demoisterizing is needed.
- Frost sensitive.
- Dependent on the place construction can be necessary.
- Packing material are for constipation by substance (> possibly sensitive 10 mg/m$^3$).
- For fragrance problems frequently pilot tests have been required value the feasibility in.

Gas scrubbers are mainly applied in the chemical - and pharmaceutical industry, detritus combustion installation, on - and transshipment of chemicals and surface treatment.

## Cyclonic Spray Scrubber

Cyclonic spray scrubbers are an air pollution control technology. They use the features of both the dry cyclone and the spray chamber to remove pollutants from gas streams.

Generally, the inlet gas enters the chamber tangentially, swirls through the chamber in a corkscrew motion, and exits. At the same time, liquid is sprayed inside the chamber. As the gas swirls around the chamber, pollutants are removed when they impact on liquid droplets, are thrown to the walls, and washed back down and out.

Cyclonic scrubbers are generally low- to medium-energy devices, with pressure drops of 4 to 25 cm (1.5 to 10 in) of water. Commercially available designs include the irrigated cyclone scrubber and the cyclonic spray scrubber.

In the irrigated cyclone, the inlet gas enters near the top of the scrubber into the water sprays. The gas is forced to swirl downward, then change directions, and return upward in a tighter spiral. The liquid droplets produced capture the pollutants, are eventually thrown to the side walls, and carried out of the collector. The "cleaned" gas leaves through the top of the chamber.

The cyclonic spray scrubber forces the inlet gas up through the chamber from a bottom tangential entry. Liquid sprayed from nozzles on a center post (manifold) is directed toward the chamber walls and through the swirling gas. As in the irrigated cyclone, liquid captures the pollutant, is forced to the walls, and washes out. The "cleaned" gas continues upward, exiting through the straightening vanes at the top of the chamber.

This type of technology is a part of the group of air pollution controls collectively referred to as wet scrubbers.

## Particulate Collection

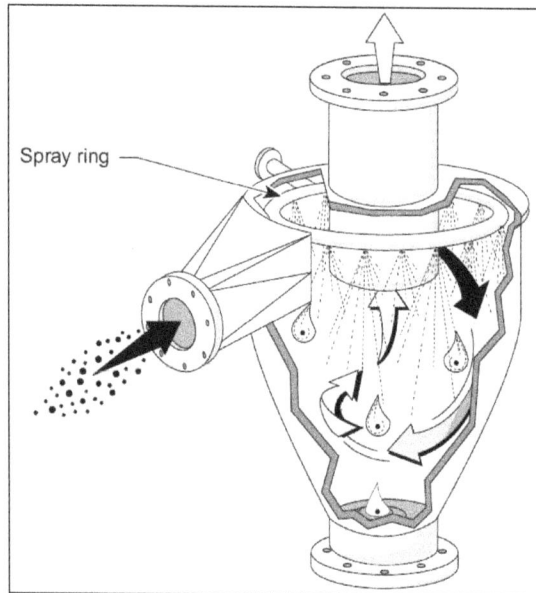

Cyclonic spray scrubber.

Cyclonic spray scrubbers are more efficient than spray towers, but not as efficient as venturi scrubbers, in removing particulate from the inlet gas stream. Particulates larger than 5 µm are generally collected by impaction with 90% efficiency. In a simple spray tower, the velocity of the particulates in the gas stream is low: 0.6 to 1.5 m/s (2 to 5 ft/s).

By introducing the inlet gas tangentially into the spray chamber, the cyclonic scrubber increases gas velocities (thus, particulate velocities) to approximately 60 to 180 m/s (200 to 600 ft/s). The velocity of the liquid spray is approximately the same in both devices. This higher particulate-to-liquid relative velocity increases particulate collection efficiency for this device over that of the spray chamber. Gas velocities of 60 to 180 m/s are equivalent to those encountered in a venturi scrubber.

However, cyclonic spray scrubbers are not as efficient as venturi scrubbers because they are not capable of producing the same degree of useful turbulence.

## Gas Collection

High gas velocities through these devices reduce the gas-liquid contact time, thus reducing absorption efficiency. Cyclonic spray scrubbers are capable of effectively removing some gases; however, they are rarely chosen when gaseous pollutant removal is the only concern.

## Maintenance Problems

The main maintenance problems with cyclonic scrubbers are nozzle plugging and corrosion or erosion of the side walls of the cyclone body. Nozzles have a tendency to plug from particulates that are in the recycled liquid and/or particulates that are in the gas stream. The best solution is to install the nozzles so that they are easily accessible for cleaning or removal.

Due to high gas velocities, erosion of the side walls of the cyclone can also be a problem. Abrasion-resistant materials may be used to protect the cyclone body, especially at the inlet.

## Baffle Spray Scrubber

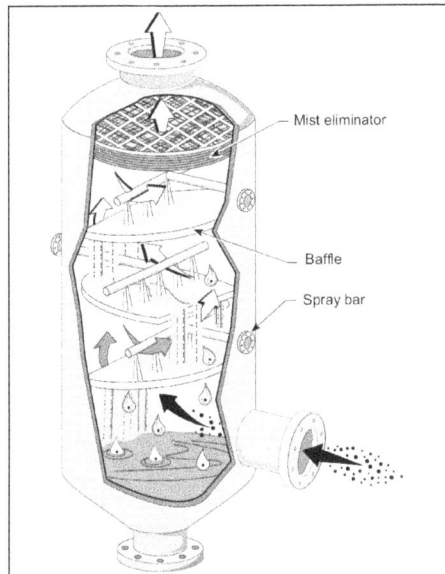
Baffle spray scrubber.

Baffle spray scrubbers are a technology for air pollution control. They are very similar to spray towers in design and operation. However, in addition to using the energy provided by the spray nozzles, baffles are added to allow the gas stream to atomize some liquid as it passes over them.

A simple baffle scrubber system is shown in figure. Liquid sprays capture pollutants and also remove collected particles from the baffles. Adding baffles slightly increases the pressure drop of the system.

This type of technology is a part of the group of air pollution controls collectively referred to as wet scrubbers. A number of wet-scrubber designs use energy from both the gas stream and liquid stream to collect pollutants. Many of these combination devices are available commercially.

A seemingly unending number of scrubber designs have been developed by changing system geometry and incorporating vanes, nozzles, and baffles.

## Particle Collection

These devices are used much the same as spray towers - to preclean or remove particles larger than 10 μm in diameter. However, they will tend to plug or corrode if particle concentration of the exhaust gas stream is high.

## Gas Collection

Even though these devices are not specifically used for gas collection, they are capable of a small amount of gas absorption because of their large wetted surface.

These devices are most commonly used as precleaners to remove large particles (>10 μm in diameter). The pressure drops across baffle scrubbers are usually low, but so are the collection efficiencies. Maintenance problems are minimal. The main problem is the buildup of solids on the baffles.

Table: Summarizes the operating characteristics of baffle spray scrubbers.

| Operating characteristics of baffle spray scrubbers | | | | | |
|---|---|---|---|---|---|
| Pollutant | Pressure drop ($\Delta p$) | Liquid-to-gas ratio (L/G) | Liquid-inlet pressure ($p_L$) | Removal efficiency | Applications |
| Gases | 2.5-7.5 cm of water | 0.13 l/m³ (1 gal/1,000 ft³) | < 100 kPa (< 15 psig) | very low | Mining operations Incineration Chemical process industry |
| Particles | 1-3 in of water | | | 10 μm diameter | |

## Thermal Oxidizer

Thermal oxidizer installed at a factory.

A thermal oxidizer (also known as thermal oxidiser, or thermal incinerator) is a process unit for air pollution control in many chemical plants that decomposes hazardous gases at a high temperature and releases them into the atmosphere.

Preassembled process unit for air pollution control, i.e., a thermal oxidizer, being installed at a work site.

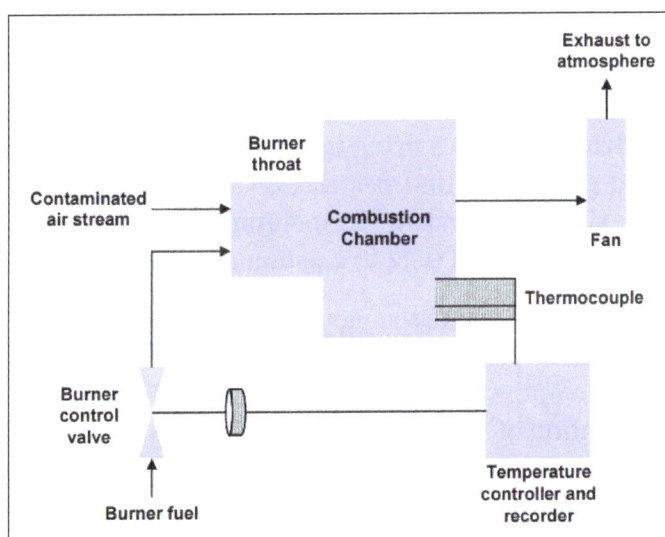

Schematic of a basic thermal oxidizer.

# Principle

Thermal oxidizers are typically used to destroy hazardous air pollutants (HAPs) and volatile organic compounds (VOCs) from industrial air streams. These pollutants are generally hydrocarbon based and when destroyed via thermal combustion they are chemically oxidized to form $CO_2$ and $H_2O$. Three main factors in designing the effective thermal oxidizers are temperature, residence time, and turbulence. The temperature needs to be high enough to ignite the waste gas. Most organic compounds ignite at the temperature between 590 °C (1,094 °F) and 650 °C (1,202 °F). To ensure near destruction of hazardous gases, most basic oxidizers are operated at much higher temperature levels. When catalyst is used, the operating temperature range may be lower. Residence time is to ensure that there is enough time for the combustion reaction to occur. The turbulence factor is the mixture of combustion air with the hazardous gases.

## Technologies

### Direct Fired Thermal Oxidizer - Afterburner

Direct-fired thermal oxidizer using landfill gas as fuel.

The simplest technology of thermal oxidation is direct-fired thermal oxidizer. A process stream with hazardous gases is introduced into a firing box through or near the burner and enough residence time is provided to get the desired destruction removal efficiency (DRE) of the VOCs. Most direct-fired thermal oxidizers operate at temperature levels between 980 °C (1,800 °F) and 1,200 °C (2,190 °F) with air flow rates of 0.24 to 24 standard cubic meters per second.

Also called afterburners in the cases where the input gases come from a process where combustion is incomplete, these systems are the least capital intensive, and can be integrated with downstream boilers and heat exchangers to optimize fuel efficiency. Thermal Oxidziers are best applied where there is a very high concentration of VOCs to act as the fuel source (instead of natural gas or oil) for complete combustion at the targeted operating temperature.

### Regenerative Thermal Oxidizer (RTO)

One of today's most widely accepted air pollution control technologies across industry is a regenerative thermal oxidizer, commonly referred to as a RTO. RTOs use a ceramic bed which is heated from a previous oxidation cycle to preheat the input gases to partially oxidize them. The preheated gases enter a combustion chamber that is heated by an external fuel source to reach the target oxidation temperature which is in the range between 760 °C (1,400 °F) and 820 °C (1,510 °F). The final temperature may be as high as 1,100 °C (2,010 °F) for applications that require maximum destruction. The air flow rates are 2.4 to 240 standard cubic meters per second.

RTOs are very versatile and extremely efficient – thermal efficiency can reach 95%. They are regularly used for abating solvent fumes, odours, etc. from a wide range of industries. Regenerative Thermal Oxidizers are ideal in a range of low to high VOC concentrations up to 10 g/m³ solvent. There are currently many types Regenerative Thermal Oxidizer on the market with the capabitlity of 99.5+% Volatile Organic Compound (VOC) oxidisation or destruction efficiency. The ceramic heat exchanger(s) in the towers can be designed for thermal efficiencies as high as 97+%.

Regenerative thermal oxidizer (RTO) that is 17000 standard cubic feet per minute, or SCFM for short.

Control center with a programmable logic controller for a RTO.

## Ventilation Air Methane Thermal Oxidizer (VAMTOX)

Ventilation air methane thermal oxidizers are used to destroy methane in the exhaust air of underground coal mine shafts. Methane is a greenhouse gas and, when oxidized via thermal combustion, is chemically altered to form $CO_2$ and $H_2O$. $CO_2$ is 25 times less potent than methane when emitted into the atmosphere with regards to global warming. Concentrations of methane in mine ventilation exhaust air of coal and trona mines are very dilute; typically below 1% and often below 0.5%. VAMTOX units have a system of valves and dampers that direct the air flow across one or more ceramic filled beds. On start-up, the system preheats by raising the temperature of the heat exchanging ceramic material in the beds at or above the auto-oxidation temperature of methane 1,000 °C (1,830 °F), at which time the preheating system is turned off and mine exhaust air is introduced. Then the methane-filled air reaches the preheated beds, releasing the heat from combustion. This heat is then transferred back to the beds, thereby maintaining the temperature at or above what is necessary to support auto-thermal operation.

## Thermal Recuperative Oxidizer

A less commonly used thermal oxidizer technology is a thermal recuperative oxidizer. Thermal recuperative oxidizers have a primary and/or secondary heat exchanger within the system. A primary heat exchanger preheats the incoming dirty air by recuperating heat from the exiting clean air. This is done by a shell and tube heat exchanger or a plate heat exchanger. As the incoming air passes on one side of the metal tube or plate, hot clean air from the combustion chamber passes on the other side of the tube or plate and heat is transferred to the incoming air through the process of conduction using the metal as the medium of heat transfer. In a secondary heat exchanger the same concept applies for heat transfer, but the air being heated by the outgoing clean process stream is being returned to another part of the plant – perhaps back to the process.

## Biomass Fired Thermal Oxidizer

Biomass, such as wood chips, can be used as the fuel for a thermal oxidizer. The biomass is then gasified and the stream with hazardous gases is mixed with the biomassgas in a firing box. Sufficient turbulence, retention time, oxygen content and temperature will ensure destruction of the VOC's. Such biomass fired thermal oxidizer has been installed at Warwick Mills, New Hampshire. The inlet concentrations are between 3000-10.000 ppm VOC. The outlet concentration of VOC are below 3 ppm, thus having a VOC destruction efficiency of 99.8%-99.9%.

## Flameless Thermal Oxidizer (FTO)

In a flameless thermal oxidizer system waste gas, ambient air, and auxiliary fuel are premixed prior to passing the combined gaseous mixture through a preheated inert ceramic media bed. Through the transfer of heat from the ceramic media to the gaseous mixture the organic compounds in the gas are oxidized to innocuous byproducts, i.e., carbon dioxide ($CO_2$) and water vapor ($H_2O$) while also releasing heat into the ceramic media bed.

The gas mixture temperature is kept below the lower flammability limit based on the percentages of each organic species present. Flameless thermal oxidizers are designed to operate safely and reliably below the composite LFL while maintaining a constant operating temperature. Waste gas streams experience multiple seconds of residence time at high temperatures leading to measured destruction removal efficiencies that exceed 99.9999%. Premixing all of the gases prior to treatment eliminates localized high temperatures which leads to thermal $NO_x$ typically below 2 ppmV. Flameless thermal oxidizer technology was originally developed at the U.S. Department of Energy to more efficiently convert energy in burners, process heaters, and other thermal systems.

## Catalytic Oxidizer

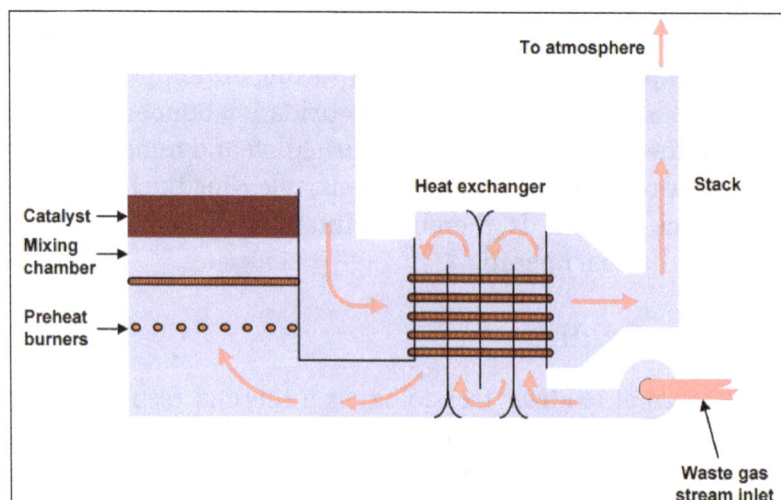

Schematic of Recuperative Catalytic Oxidizer.

Catalytic oxidizer (also known as catalytic incinerator) is another category of oxidation systems that is similar to typical thermal oxidizers, but the catalytic oxidizers use a catalyst to promote the oxidation. Catalytic oxidation occurs through a chemical reaction between the VOC hydrocarbon molecules and a precious-metal catalyst bed that is internal to the oxidizer system. A catalyst is a substance that is used to accelerate the rate of a chemical reaction, allowing the reaction to occur in a normal temperature range between 340 °C (644 °F) and 540 °C (1,004 °F).

## Regenerative Catalytic Oxidizer (RCO)

The catalyst can be used in a Regenerative Thermal Oxidizer (RTO) to allow lower operating temperatures. This is also called Regenerative Catalytic Oxidizer or RCO. For example, the thermal ignition temperature of carbon monoxide is normally 609 °C (1,128 °F). By utilizing a suitable oxidation catalyst, the ignition temperature can be reduced to around 200 °C (392 °F). This can result in lower operating costs than a RTO. Most systems operate within the 260 °C (500 °F) to 1,000 °C (1,830 °F) degree range. Some systems are designed to operate both as RCOs and RTOs. When these systems are used special design considerations are utilized to reduce the probability of overheating (dilution of inlet gas or recycling), as these high temperatures would deactivate the catalyst, e.g. by sintering of the active material.

## Recuperative Catalytic Oxidizer

Catalytic oxidizers can also be in the form of recuperative heat recovery to reduce the fuel requirement. In this form of heat recovery, the hot exhaust gases from the oxidizer pass through an heat exchanger to heat the new incoming air to the oxidizer.

# Reduction of Air Pollution by Combustion Processes

Among human activities, those in which fuel combustion processes intervene are those who contaminate the atmosphere greatly.

The combustion process is a process of rapid oxidation, followed by light phenomena and the release of large amounts of energy, able to maintain it at high temperatures. Compared with slow oxidation processes, it is characteristic to the combustion process sudden acceleration of the reaction rate to achieve theoretically infinite values. This applies, for example, to the stoichiometric mixture of methane - oxygen heated to a temperature of 560 °C in a sealed container. Heating the same mixture to a temperature of only 200 °C, result in a slow oxidation process, which produces methanol, formic acid, formaldehyde, carbon monoxide and carbon dioxide gas, with an overall response rate with an evolution with measured values up to a maximum, after which rate value decreases with the depletion of reagents. In everyday life we encounter slow oxidation processes at every step. Thus, minerals are subject to slow oxidation process which occurs at ambient temperature by consumption of oxygen from atmospheric air, with production of oxides in a state of maximum stability. Such a process is carbon steel corrosion under the action of atmospheric oxygen at ambient temperatures, which is transformed first into ferrous oxide (FeO) and then in a more stable substance, ferric oxide ($Fe_2O_3$). Also, living organisms consume oxygen in the atmosphere, at room temperature to oxidize nutrients over a slow but very complex process. In both examples above, as in any oxidation process, there are necessary two substances: the oxidant, which has the ability to quickly combine with the substance subject to oxidation, respectively, the substance that is oxidized, called fuel.

The transformation of chemical energy of fossil fuels in forms of energy directly useable, primarily mechanical energy, electrical energy and heat energy, is practically done only by means of combustion. In the production of electrical and heat energy, are consumed by burning, at present, 87% fossil fuels, the remainder being nuclear energy and regenerative energy (hydraulic, wind energy, solar, geothermal and marine) 6%, respectively 7%. At this consumption of the fossil fuels the consumption to produce mechanical energy in transports and the technological consumption, e.g. consumption of coal to produce metallurgical coke and for injection in blast furnaces, is added.

In 2008, according to the world consumption of fuels was 12 300 million toe (tons oil equivalent), of which 30 660 million barrels of oil and 3100 billion cubic meters of natural gas. For the year 2035, according to the script "new scenario" of the total consumption will be increased up to16 700 million toe, of which 36 135 million barrels of oil and 4500 billion cubic meters of natural gas. These significant increases in world consumption of fuels results in conditions that anticipate an increase in the share of nuclear energy to 8% and regenerative energy to 14% in electrical and heat energy production.

In the fuel combustion processes the maximum release of energy is obtained with the complete combustion with, minimum excess of air, closer to stoichiometric combustion, which generally leads to the exclusive formation of $CO_2$, $H_2O$, $SO_2$ and $N_2$ gases, respectively $O_2$ and $N_2$ of the air excess. In practice, combustion processes slide away more or less from this ideal, both due to the complex structure of used fuels, such as coals and inferior oils, also due to imperfection of the combustion installations. Thus are formed unburned substances as solid particles, rich in carbon, which in most cases are accompanied by unburned gases. Unburned substances discharged into the atmosphere with the combustion gases, lead to heat energy loss, which leads to a decrease of the thermal efficiency of the process but at the same time, they have polluting action.

Sources of air pollution from burning fuels are classified as follows: stationary sources, mainly consisting of boilers, furnaces and gas turbines and mobile sources consisting of transport means, whose internal combustion engines are diesel (diesel-consuming, also called compression-ignition engines) and spark ignition engines, consuming gasoline. Lately there are used as stationary sources in heat energy and electricity production in cogeneration and internal combustion engines. Classification of stationary sources is done by industry type:

- Energy Industry (heat energy and electricity production, petroleum refining, manufacture of solid fuel and other industries);

- Manufacturing Industry and construction (iron and steel, chemicals, pulp, paper and printing, food processing, beverage and tobacco), called the EEA Industry (energy) sector. Also in EEA (European Environment Agency) publications mobile sources are classified in road transport and other transport.

Especially in large steam boilers, through the burning of coals and oils, solid particles driven by exhaust gases may contain non-ferrous metals (Pb, Zn, Ni, Cd, Sn etc.) and metalloids(F and As).

In diesel engines, through incomplete combustion the fumes can appear white, blue respectively, which consist of a suspension in gases of liquid particles of unburned fuel or partially oxidized. Particles with diameter over 1 µm are found in the white smoke and particles with a diameter of about 0.5 µm appear in the blue smoke.

Black smoke from diesel engines contains solid carbonaceous particles, with average diameters of 1 μm. Some experts consider that solid primary particles ranging in size from 0.01 to 0.05 μm, during the expansion and exhaustion of burning gases, find suited conditions to aggregate in soot particles with dimensions up to 0.6 μm occupying a specific surface of about 200 m²/g. Experimentally it was found that these particles are formed under certain specific conditions, of: 20-90% amorphous carbon, 2-5% ash and 10-80% carcinogenic hydrocarbons.

In order to diminish air pollution by combustion processes, the following solutions are recommended:

- Reducing the pollutant emission at origin, at the producing sources (primary procedures) by changing-replacing the combustion processes (use of renewable energy, production reducing, increase of production/technological lines efficiency, replace of production technologies), by using pretreatment procedures of fuels and oxidizer substances (changing fuels and oxidizers, improving the fuels characteristics by processing, preparation and use of additives) and by using improvement – changing procedures of combustion processes, both in case of combustion installations - especially the burners and burning chambers with reference to their construction (mixed stages, flue gases recirculation, auto-carburizing, steam or water spraying) and in case of combustion adjustment (control of excess air, burners out of service, adjustment of combustion air temperature);

- Collecting and/or reducing the pollutants resulted from flue gasses before their suppression into the air by using the so-called secondary pollution reducing techniques or flue gases treatment (catalytic and non catalytic reactors, active carbon processes, SNOX and DESONOX processes, electrostatic precipitators, bag filters, scrubbers etc).

## Formation and Reduction Pollutants in Combustion Processes

## Carbon Monoxide

The emission of carbon monoxide (CO) pollutant in combustion processes is produced when some conditions occur which lead to incomplete burning of fuels that contain carbon. Among heavy users of the most polluting fossil fuels, through emission of CO, are the means of transportation driven by spark ignition engines and compression ignition engines, and the lesser CO polluting are power plants.

If oxygen is missing during the combustion, compared to stoichiometric requirements, in an area of a fuel-air mixture, it will result CO even at a sufficiently high temperature. The possibility of CO subsequent oxidation, when the contribution of necessary oxygen to a lower temperature appears, is slower than forming CO.

In boilers and in furnaces, the presence of CO in the exhausted flue gases highlight essentially a disorder of the combustion equipment: burner, sprayer, combustion air blow installation, etc. Carbon oxide appears usually in the flue gases discharged from old, rudimentary combustion plants and is almost absent in the exhaust gases from modern industrial combustion plants of high-capacity. Also, at these facilities is taken into account that reducing the excess of air, which reduces the emissions of $NO_x$, can increase CO emission. As a result an optimal value for the air excess coefficient is established.

Ensuring complete combustion is a measure considered to be BAT (best available techniques), which ensure the maintenance of the CO emission levels below 100 mg/Nm$^3$, for 15% $O_2$ (for internal combustion stationary engines and gas turbines) and in boilers and furnaces, for 3% $O_2$, to run on liquid and gaseous fuels and 6% $O_2$ to run on solid fuels.

For mobile sources, which are the biggest pollutants with CO, as reduction techniques were developed, among others, multi-point injection engines that work at high pressures and oxidation catalysts.

The amount of CO emitted by transportation in European Union countries (EU-27) represents 89% of the total CO emissions at their level. Due to equipping vehicles with better catalysts and application in EU-27 of new regulations that produced engines up to Euro-6, CO emissions diminished, in 2008, compared to 1990, with 71%.

## Sulfur Oxides

Beside some gaseous fuels, the other industrial fossil fuels generally contain sulfur S. For engines, diesel has a low sulfur content, which is null to petrol diesel prompts. Sulfur content in diesel determines $SO_2$ polluting emissions, which on diesel engines represent about 10% of the polluting emissions of these engines, but are being endangered as a result of production, lately, of sulfur-free diesel fuels.

Maximum sulfur content from main fossil fuels is regulated, e.g. France, S 4% for industrial liquid fuels and S 1% for solid fuels.

Primarily coals and oils, which are used in industrial plants to produce electricity and heat, are responsible for environmental pollution by sulfur oxides, strong acidity of air and acid rains. Sulfur dioxide tendency is to reduce the "greenhouse effect" by "albedo" effect.

Crude oils contain variable quantities of sulfur depending on their origin. During refining operations, sulfur tends to gather in waste fractions which can be an important element to the industrial liquid fuels. The difficulty to reduce the sulfur content in crude oils and petroleum residues consist especially in their content of asfaltene and metals which lead to deactivation of catalysts. Researches are made in this respect. At a large scale of solid fuels the sulfur content is close to the values for heavy liquid fuels. Also, desulphurization of solid fuels with high content of sulfur is a major activity, that develops both through practical achievements and through researches.

Discharged in the atmosphere, $SO_2$ react in proportion of 1-2 ‰/h with oxygen under the action of solar ultraviolet radiation resulting $SO_3$ which combines with water vapor in the atmosphere to form sulfuric acid. In very wet days, with fog, the degree of conversion to sulfuric acid is up to 15.7%.

Also, starting from $SO_2$ gas, sulfate aerosols forms but they can be washed by the rain which becomes acid. To ensure a satisfactory dilution and dispersion of the $SO_2$ gas in the atmosphere, the combustion gases must be evacuated from the funnel with an adequate speed and to a convenient height from the ground. Thus the large thermoelectric oil plants funnels exceed 200m height frequently. Because of this the pollution due to pollutants produced by sulfur in the fuel may occur at great distances from where they were generated.

Generally, the growth of sulfur dioxide quantity in the atmosphere has the effect of increasing the overall morbidity rate of the population. At concentrations of sulfur dioxide in the cities air over 0.046 ppm (annual average) lead to an increasing frequency of respiratory diseases, at concentrations of 0.52 ppm in the presence of other solid particles, overall mortality is increased. Sulfur dioxide and sulfuric acid aggravate the respiratory system of animals, but this is much more toxic than $SO_2$, its action being dependent on the size of aerosols. In a series of countries, particularly those located in northern area, the specific flora is affected even by the traces of sulfur dioxide, moss and lichens around large cities from Europe and America, disappear. Sulfur dioxide absorbed by plants leads to acute and chronic effects. Acute effects, that follow high concentrations and the relatively short exposures, manifest by changing the color (to yellow-ivory or red-brown) of damaged tissues. Chronic effects, results of prolonged exposure to low concentrations, manifest by perpetual yellowing of the plant foliage following the alteration of the chlorophyll production mechanism. It seems that, to the acute effects, the plant defend itself transforming sulfur dioxide into sulfuric acid and then in sulfates that are deposited in certain portions of tissues, but soon the balance of sulfates is disturbed and sulfuric acid appears in the plant cellular system which attacks its cells. Chronic effects are due to sulfate aggregations. Sulfur dioxide in small concentration (on the order of the size of 100 - 500 $g/m^3$) react synergistically with both ozone and nitrogen dioxide, which affects the plants tissue, especially as these are more sensitive.

Sulfur dioxide and sulfuric acid formed have very corrosive action on metals, constructions, leather, paper and textiles, especially if relative humidity of air exceeds 70%. Attack of construction materials by $SO_2$ is due to next mechanism: $SO_2$ in the atmosphere prompts a condensation of water vapors as fog even when their partial pressure is lower than the saturation pressure at that temperature, due to the fact that $SO_2$ form with water a solution which a lower vapor pressure than pure water, slip by easily as fog state; the phenomenon can be facilitated by existing aerosols as particles in the smoke, which forms condensation nuclei. In soft drops of fog with $SO_2$ is obtained sulfurous acid and its oxidation to sulfuric acid. Reducing character of $SO_2$ manifests also by changing the color of some paint pigments. As such it is affected lead white. Calcium carbonate, a constituent of many construction materials, some of which serve to works of arts, is transformed into calcium sulfate. Destructive effect is accelerated by the acid rains.

In 1983 the first symptoms of forests etching have appeared in Western Europe, due to acid rains which gather both acid pollution and photooxidant pollution. It has been established that acidity in the atmosphere, most often originates in a ratio of 2/3 due to emissions of sulfur oxides, less than 1/3 because of $NO_x$, and the rest due to pollutants as fluorine or chlorine. Photooxidant pollution is due transforming $NO_x$ in the presence of hydrocarbons and solar radiation.

Beside the negative effects outlined above, the presence of sulfur oxides in the air space reduces the visibility when the fog photochemical processes occurring and determines the appearance of an unpleasant smell.

In the atmosphere, oxidation of $SO_2$ to $SO_3$ takes place mainly through a photochemical process, although there are some opinions that it could take place catalytic processes also. During these reactions the presence of other pollutants such as nitrogen oxides or metal oxides may play an important role to accelerate the reactions. But there are occuring other factors such as concentration, residence time in the atmosphere, temperature, humidity, intensity and spectral distribution of radiations. In boilers furnaces sulfur trioxide, or sulfuric anhydride $SO_3$, derives from the oxidation sulfur dioxide. Actually, in the combustion gases is found 3-5% of the initial sulfur contained in the fuel, as $SO_3$.

Under certain circumstances $SO_2$ combines with water vapors, giving sulfurous acid gas $H_2SO_3$ and $SO_3$ combines still with water vapors, giving gaseous sulfuric acid $H_2SO_4$. Acid vapors condensation on cold walls attack metals. Beside corrosion, due to $SO_4H_2$ vapors condensation, especially in thermal uninsulated funnels, it is favored the agglomeration of solid particles of soot and flying coke, who create acid rain too.

The fight against acid corrosion and pollution is done in three ways:

- Preventing the forming of anhydrides, especially $SO_3$;

- Neutralizing the formation of acids;

- Preventing the condensation of formed acid vapors.

Avoiding actual forming of anhydrides, means burning fuels with a low sulfur content. Heavy petroleum desulphurization is not generalized to an industrial scale today because of the high prices required. Fuels with low sulfur content can be subjected to industrial operations only when sulfur compounds are extremely harmful. Sometimes it is possible to have them burned out at low excesses of air, thus forming reduced $SO_2$ and $SO_3$ quantities. Limiting the excess of air at maximum 3% is difficult to do because:

- The burner must ensure the right mix between combustion air and fuel over the whole tuning;

- Industrial boilers being equipped with numerous burners, each burner must have its own system of adjustment and control, which raise the cost of the related equipment. So the solution is applied only to very large boilers from thermoelectric power plants, which have numerous personnel for maintenance and exploitation.

Spraying silica allows to stop the catalytic action of deposits that form on the heating and overheating pipes, which reduces the amount of formed $SO_3$. Frequent cleaning of depositions of solid particles completes the action mentioned before, but does not reduce the formation of $SO_3$ from the flame. $SO_3$ action may be partially neutralized by injection into the combustion products of calcium carbonate, zinc oxide, dolomite and of ammonia, which allows some boilers to reduce significantly the corrosion of air preheaters. Injection technique is sensible and it must be assured a good distribution of the substances mentioned above, avoiding obstruction of the flue gas flow and solid particle entrainment in atmosphere. When using ammonia, neutralization does not stop at the stage of acid sulfate of ammonia, but it reaches the stage of neutral sulfate.

Primary procedures for reduction of $SO_2$ pollutant emissions make the desulphurization, reducing the content of $SO_2$ in the flue gas by injecting absorbent substances sprayed into the furnace, such as limestone, lime or dolomite. Desulphurization combustion can be seen as a primary technique for major reduction of $SO_2$, applied to certain types of boilers. For the first achieved version, when the flame is obtained from a conventional burner, the U.S. has developed an extensive research program LIMB (Limestone Injection in Multistage Burners) with several forms of application.

Desulphurization efficiency depends particularly on the magnitude of the temperatures in the furnace, but meritorious performances are obtained using high consumption of absorbent.

For the second achieved version, the combustion is developing in dense or circulating fluidized bed. Such techniques can achieve efficiencies of combustion and desulphurization with high values having a consumption of absorbent lower than the first version. It should be noted that circulating bed technique is superior as efficiency than the dense bed technique.

Dense fluidized bed is characterized by the flow speed, generally reduced, where the solid particles move relatively slow and require a moderate recirculation. Increasing the speed of flow, the volume occupied by the layer will increase, as the movement speed of solid particles, which will require an intense recirculation of particles realizing the circulating fluidized bed. These particles are of two types:

- A kind of incompletely burned fuel, because the coal particles introduced into the furnace having a high temperature generate gaseous volatile mater and carbon tailing which require a much longer combustion time than the combustion gases;

- Absorbent material, which is mostly limestone, which has three forms: $CaCO_3$, $CaO$ and $CaSO_4$.

Because the system operates in stationary state, the continuously exhausted gases will have to contain very small quantities of combustible substances to obtain the highest possible energy efficiency. Also, the exhaust gases will contain $CaCO_3$ and $CaO$ as well as a very high content of $CaSO_4$ for a good desulphurization. Conventional circulating fluidized bed boilers realize the burning and desulphurization in the same furnace known as reactor that is fed simultaneously with fuel and absorbent material. The differences between these boilers refer to the ways of extracting produced energy and the means of control operation of variation with load. IFP, Lardet-Babcock and Cecar made a boiler called self desulphurization to which combustion and desulphurization are produced in different rooms. This boiler has a circulating fluidized bed furnace which is fed with combustion gases, having the temperature of 500-850 °C, instead of air, as conventional boilers are operating. Hot combustion gases, rich in $SO_2$, from the furnace, are routed through the loop of circulation of solid particles in suspension, where is added limestone. This loop does not have exchange surfaces for useful transfer of heat, so it is protected from the effects of the sulfuric corrosion. The combustion gases thus sweetened are directed to the heat exchange surfaces of a exhaust-heat boiler. Whatever the system, in principle, all types of boilers with circulating layer are operating with relatively soft particles of limestone, of a few hundred microns, rather than a few millimeters bed as the dense bed boilers user. We mention that circulating bed systems allow a better use of calcium than dense systems.

Although there are few studies of the kinetics for desulphurization in circulating bed, there are global and fragmentary results often achieved on industrial boilers where the degree of desulphurization was measured in given operational conditions, knowing the nature of absorbent and the fuel burned and the Ca/C ratio, respectively.

Secondary procedures for reducing pollutant emissions of $SO_2$ use desulphurization plants of exhaust gases. The most used are dry FGD (flue gas desulphurization) plants or wet FGD plants and to a lesser extent plants combined, for desulphurization and $NO_x$ reduction of exhaust gases, as DESONOX, which are regarded as techniques associated with BAT. FGD plants with lime or limestone wet scrubber have a rate of $SO_2$ reduction by 92-98%, those who dry scrubbers, with spray, by 85-92%.

Depending on the type of installation of boilers, old or (new), as its power, by applying the BAT techniques, the emission levels of $SO_2$, in mg/Nm³, are limited, e.g. for liquid fuel (calculated at 3% $O_2$) to 350 for powers under 100MWt, to 250 for powers under 300MWt and to 200 (150) for powers over 300MWt.

The amount of $SO_2$ emitted in producing electricity and heat sector in European Union countries (EU-27) represents 69% of the total $SO_2$ emissions at their level (energy using industry sector is 10.2% and transportation 3.8%). Due to replacing solid and liquid fuels with high sulfur content with natural gases, the use of technologies of flue gas desulphurization and application in EU-27 of the new regulations which limited the sulfur content of some liquid fuels, the emissions of $SO_2$ have decreased from approx. 26000 kt, in 1990, by 71%, to approx. 7500 kt in 2007.

## Nitrogen Oxides

Among heavy users of the most polluting fossil fuels, through the $NO_x$ emissions, are the means of transportation driven by spark ignition engines and compression ignition engines, followed by thermoelectric power plants and then by boilers and furnaces in the sector of energy use in Industry.

$NO_x$ emission depends on the residence time of the molecules in the flame, most of $NO_x$ is formed in the second part of flame development, where the temperature is sufficiently high. Usage of the excess air raise the $NO_x$ equilibrium value, and the time needed to achieve balance is net superior to the stationary time of the molecules in the furnace, which explains the experimental values of $NO_x$ lower than the theoretical front.

There are three possible mechanisms of formation the $NO_x$:

- Thermal $NO_x$ from using the nitrogen from the combustion air;

- Fuel $NO_x$, by conversion of chemically bound nitrogen in the fuel;

- Prompt $NO_x$.

$NO_x$ forming synthesis is detailed schematically as follows:

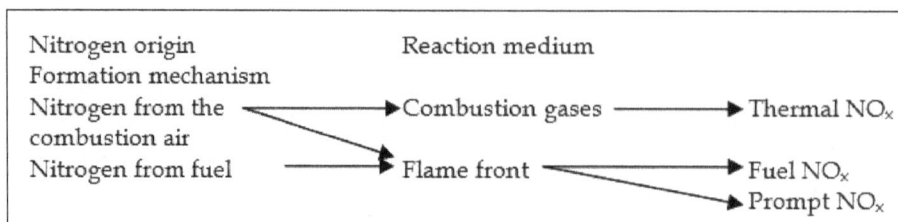

The Zeldovich chain reaction pattern shows the exponential dependence for temperature of the emission of $NO_x$, until the temperature threshold is reached. The emission is directly proportional with the residence time in the furnace of the molecules and with the oxygen square concentration.

The temperature is the strongest factor to influence the $NO_x$ emission, exceeding significantly the influence of the oxygen concentration and the stationary time of the molecules. Stepped combustion is mainly intended to avoid forming high temperatures for reactants before reaching equilibrium, thus nitrogen oxides overall emission to be lower.

The wrong placement of the burners may determine through the jets interaction phenomenon local increases in temperature, which can lead to increased emissions of nitrogen oxides. Fuel nitrogen is present as organic compounds in the amines (NH and NC) and cyanides (CN) family.

These compounds react in two directions by:

- Reaction with substances that contain oxygen, forming NO.

- Reaction with substances containing nitrogen, forming molecular nitrogen.

Therefore not all the fuel nitrogen from combustion passes in $NO_x$.

Thermal NO formation occurs in the flame in the post-reaction area, according to distinct mechanisms corresponding to mixture of fuel with poor or rich air.

The main factors that influence the development of thermal NO production reactions are: the atomic oxygen concentration, the residence time of molecules and the temperature of the furnace, with higher values at 1300 °C, exert the strongest influence.

The reaction of molecular nitrogen with atomic oxygen is the slowest. Dissociation of molecular oxygen at normal combustion temperatures is insignificant. Thus, for example at 2000 K, atomic oxygen is formed having a concentration below 10 ppm.

NO formation from fuel nitrogen occurs in the flame after a complex mechanism, partially unknown. It is known that first CN radicals are formed, whose evolution in the presence of oxygen leads to the formation of NO. The main factors that influence the forming of NO in this case are the nitrogen content of the fuel, oxygen concentration in the fuel, residence time and the flame temperature.

NO formation from the nitrogen contained in the fuel flows a little faster than the formation of thermal NO, but is considerably slower than the prompt NO formation.

$NO_2$ formation occurs in the exhaust ducts of combustion gases, funnel and free atmosphere at temperatures below 650 °C, the main factors being: the temperature value, molecular oxygen $O_2$ concentration, residence time, air pollution and solar radiation.

Primary procedures for reducing pollutant emissions of $NO_x$ have known three generations in development.

- First generation consisting in reducing the temperature of the combustion air preheating (RAP), flue gas recirculation application in the furnace (FGR), use of low excess of combustion air (LEA) and burners-out-of-service (BOOS);

- The second generation consisting in production of low $NO_x$ burners 1 (LNB) or air staging at burner, flue gas recirculation at burner (FGR) and over fire air (OFA);

- Third generation consisting in production of low $NO_x$ burners 2 (LNB) or air and fuel - staging at burner and in furnace $NO_x$ reduction (IFNR) as well as burners improvement for $NO_x$ decreased by increasing the internal flue gas recirculation.

In addition to these techniques there can be mentioned:

- Emulsification of oil with water;

- Injection of aerosols;

- Water-steam injection;

- Addition of fuel, combustion air, or spraying additives in flame.

$NO_x$ emission depends on the size of volume of the furnace in which take place the development of flames produced by burners. Large furnaces allow the existence of higher temperatures in and after the area occupied by the flame, so the $NO_x$ emission will increase, and the phenomenon is amplified by increasing the residence time. As a result, thermal charges of the volume and the cross section of the furnace represent important criteria for assessing the emissions of $NO_x$.

In large furnaces, which burn gaseous fuels, the $NO_x$ emission is reduced by lowering the preheating temperature of the combustion air. Using this technique to oil burning is not recommended because it increases the percentage of unburned, especially soot production, as a result of lowering the burning temperature.

The amount of preheating temperature of the combustion air must have an optimum determined by the combustion to be more completely, $NO_x$ emission below the normalized limit and high thermal efficiency for the heat production plant.

For industrial furnaces, where high temperatures in the furnace are required by the technological process, the amount of preheated temperature of the combustion air is dependent on this need taking into account, firstly, the type of the fuel used. Primary reduction of $NO_x$ emission is done especially by removing the temperature peaks that are achieved mainly by the combustion gases recirculation.

Flue gases recirculation in the combustion air represent an efficient method of reduction of $NO_x$ emissions for fuels and furnaces that allow development of high temperatures in the flame. The researches have shown that the most effective method of decreasing the levels of $NO_x$ is the recirculation gases mixture with the primary air or gaseous fuel, because in this case it acts on the maximum of temperature of the flame core. Thus the burning rate is reduced, which increase the length of the flame and reduces the $NO_x$ content. As result there are necessary constructive changes to the burners and limiting the quantity of recirculation gases, so the complete combustion to be assured.

But it is noticed that by increasing the recirculation degree, the flue gas CO content is increasing, while the $NO_x$ content is decreasing. Reducing the temperature of the flame decrease the useful transfer of the heat by convection, because of the increase of the gases flow from the combustion of fuel with recirculation flue gas flow.

To the Hrenox process, recirculation gases mixture with some cold combustion air is introduced into the burner separately from the rest of the combustion air that is preheated. For a high degree of flue gas recirculation of 15% and a proportion of cold air up to maximum 10%, the $NO_x$ emission was below 200 mg/m³N for the combustion of gaseous fuels and below 250 mg/m³N for the combustion of high quality liquid industrial fuels.

Another way to reduce $NO_x$ emissions is in using tertiary air injected through special slits, located above or below them.

The burners will work with decreased excess of air, the combustion being done by tertiary air which represent 15-30% of total combustion air. Often the application of this combustion technique is related to the existence of large furnaces, equipped with several rows of burners located at different heights, where only to the upper burners it may be introduced by blowing tertiary air.

Also, to the individual burners of natural gas, or oil, the burning in steps is a safe technique reduce the $NO_x$ emissions. Combustion by-steps can be obtained by introducing fuel gradually by FS (air- fuel staged) process. The fuel added outside the primary burner forms a more understoichiometric flame after which, in the second stage $NO_x$ is reduced by $NH_3$, HCN and CO radicals in $N_2$. Optimal proportion of the secondary fuel from total fuel consumption is 20-30%. The third stage follows when the process completes when the final combustion with reduced $NO_x$ is done.

Insertion by-steps of the combustion air or fuel, reduces the $NO_x$ emissions, but this process must be implemented by optimal constructive solutions, whereas the $NO_x$ reduction processes are contrary to those of decreasing the CO and partially oxidized hydrocarbons.

The time required for combustion and reduction being relatively high, the FS process is recommended to apply to large furnaces. At the oil combustion, the FS process application may produce significant soot quantities in the primary flame, so it must be applied judiciously in case of combined oil-gas burners.

INFR combustion by-steps process implies the fuel injection in the furnace above the main combustion zone thereby causing a secondary under-stoichiometric area, after which it is added downstream the secondary air which completes the combustion. As a result, the combustion is divided into three zones. Hydro-carbonate radicals, formed in the second zone, occur to a temperature of over 1200 °C in the reduction atmosphere, reacting with nitrogen oxides produced in the main combustion zone so the $N_2$ and other components to be formed. In practice, reducing $NO_x$ emissions by spraying water is rarely used, in large furnaces and only with superior fuels that form relatively high combustion temperatures and who have virtually no sulfur in the composition.

In the case of heat engines, primary measurements to reduce $NO_x$ emissions refer to optimizing the combustion chamber, the injection or carburetion systems, engine operation with an appropriate adjustment, using recirculation of exhaust gases, to which it is added the use of fuels treated by different methods (desulphurization, emulsifying with water, ultrasonic treatment).

Most of the researches conducted at the boilers to reduce the emissions of $NO_x$ produced by burners referred to the oil - natural gas mixed burners to which the constructive solutions adopted are independent of fuel that they use. Primarily it is used the combustion by-steps and the combustion gases internal recirculation, as the ASR (Axial Stage Return Flow) burner is done. Part of the combustion air flow passes through the central tubing of the burner as primary air. The secondary air is repressed from the burner box through pipes equidistant distributed on a peripheric circumference, each pipe having two outputs for secondary air I and secondary air II, the supply being performed by a separate cold air fan. To delay the combustion with primary air, due to the ejection caused by axial repression with higher speed of secondary air, flue gases are absorbed from the furnace through a annular transversal section, forming a separating layer between the primary step and the secondary step. In this burner there were integrated essential components of techniques to reduce $NO_x$ emissions:

- Supply with combustion air in three stages;

- Aspiration of the flue gas from the furnace ensuring internal recirculation;

- Separation of primary air from secondary air through a stream of combustion inert gases that reduces the burning rate;

- Liquid and gaseous fuels flows are adjustable allowing the control of the combustion processes to conditions of reduced $NO_x$.

Very low $NO_x$ emissions were obtained by using the ASR burner alone and by applying additional techniques such as the addition of tertiary air, external flue gas recirculation and special spray nozzles for liquid fuel. To the combustion of oils with high sulfur content, according to the quality of these fuels, for entry into the solid particles and sulfur oxides emissions standards it is required the application of post-combustion control techniques that reside in the use of desulphurization systems, filters and cyclones.

For the furnaces of boilers burning solid fuel with low volatile content and a lot of bituminous ash having the liquid discharge of ash, $NO_x$ emissions level is relatively high. Since it is necessary to maintain a minimum value of the combustion gases temperature required for the liquids discharge of ash, $NO_x$ emission control techniques are restricted in operation. Cyclone furnaces with second chamber downstream offer very good conditions for combustion by-steps. When inserted into the cyclone furnace the required air for combustion there are obtained high temperatures for oxygen excess, so the $NO_x$ emissions have high values to 1500-1800 mg/Nm³. To the combustion by-steps the procedure is as follows:

- In the cyclone furnace is inserted about 80% of stoichiometric air;

- It is injected in the secondary chamber, as secondary and tertiary air, the remaining air required for fuel combustion. The optimal conditions for injection with tertiary air are met when the injection nozzles are designed and located so to provide the intimate mixture of the flue gas from the cyclone furnace and air. Respecting the above conditions, it was obtained a reduction in $NO_x$ emissions of 30-40% depending on the operating mode. Further reduction in $NO_x$ emissions was achieved by applying the external flue gas recirculation.

For mixed gas-oil burners the reduction of $NO_x$ emissions is done by the swirling the combustion air to a turbulence degree of n<3 to prevent high temperature peaks as a result of excessive shortening of the flame.

Intensifying the internal recirculation of combustion gases is done with repressed secondary air with great speed through nozzles located at the end of the entry into splay, creating an important vacuum that absorb primary combustion products. Combustion air admission is done best in three steps to avoid local temperature increase to high values. It is recommended that the primary air to represent 55-60% of the total combustion air. Oil spraying is done especially with slightly over-heated steam using Y-type injectors.

To reduce $NO_x$ emissions and pollutants in general, it can be used the self-carburizing process for both boilers and furnaces.

A low burner of original conception, with solid carbon particles formation in flame, is fitted on three condensation boilers for steam and hot water. The main burner elements provide the formation of annular jet of combustion air mixed with natural gas and of central fuel jet, which, by thermal cracking, forms subsequently particles of solid carbon. These particles lead to increase of flame emissivity. Having in view the $NO_x$ emission, the characteristics of this burner are the following:

- Use of natural gas as fuel;

- The solid carbon particles formation in flame;

- Three stages mixing of the fuel in the combustion air;

- The swirling motion of the ventilated combustion air.

The low $NO_x$ burner for boilers demonstrated excellent $NO_x$ and CO performance, producing emission levels below 60 ppm for CO(1.05 air ratio at nominal power and 1.09 air ratio at minimum power) and below 50 ppm for $NO_x$ (1.03 air ratio at nominal power and 1.09 air ratio at minimum power), operating with natural gas.

Another low $NO_x$ burner of original conception, with solid carbon particles formation in flame, is fitted on forge and treatment furnaces. The main constructive elements of this low $NO_x$ recuperative burner are: a ceramic quarl, a zone for the natural gas-ventilated air mixture, a zone for fuel atomization by collision with the compressed air, an air preheater and an ejector for exhaust gases. The ceramic quarl burner has a cylindrical combustion chamber and in the peripherical zone a few cylindrical channels for exhaust gases entrances. Between combustion chamber and cylindrical channels are another channels for exhaust gases recirculation. The zone for the natural gas-ventilated combustion air includes three concentric pipes: the central pipe with different orifices and nozzles for natural gas and two pipes for divided combustion air in primary, secondary and tertiary air. The air preheater ensures the heat exchange from exhaust gases to combustion air. Having in view the $NO_x$ emission, the characteristics of this burner are the following:

- Use of natural gas or heavy fuel;

- Three stages mixing of the fuel with the combustion air;

- The solid carbon particles formation in flame;

- The exhaust gases recirculation.

The low $NO_x$ recuperative burner demonstrated excellent $NO_x$ and CO performance, producing emission levels below 30 ppm for CO (1.03 [1.08] air ratio, and 700 °C furnace temperature) and below 120 ppm for $NO_x$ (1.30 [1.35] air ratio, and 1200 °C furnace temperature), operating with natural gas [heavy fuel].

In conclusion, the solutions for upgrading the combustion plants for fossil fuels for boilers and furnaces can be grouped into:

- The completion of the measurement, control and automation equipment and establishing optimal operating parameters;

- The development of new types of burners;

- The development of new combustion installations with high complexity.

The following can be a part of the first type of solutions:

- Operation with optimal values of the coefficient of combustion excess;

- Operation with optimal values of the preheating temperature of the combustion air and the gaseous fuel, like bfg (blast furnace gas) that is used;

- Operation with an optimal number of burners with a given unitary capacity.

The second group refers to making of:

- Burners with the insertion by-steps of combustion air or fuel;

- Burners with internal recirculation of flue gas;

- Burners having gradual combustion.

In the last group of solutions there are:

- Making of combustion plants with external recirculation;

- Making of combustion plants with water-fuel liquid emulsions or additived aerosols for fluid ionization in the furnace;

- Replacing of the solid fuels with liquid or gaseous fuels that are less pollutant. By applying primary techniques, the rate of reduction of $NO_x$ emissions is 10-44% for operating with reduced air excess, 10-70% at the insertion by-steps of the fuel or oxidizer, 20-50% at flue gas recirculation, 20-30% at low preheating of the combustion air and 25-60% to use low $NO_x$ burners.

Secondary procedures for reducing $NO_x$ pollutant emissions use flue gases treatment plants. The most commonly used installations are SNCR (selective non-catalytic reduction) and SCR (selective catalytic reduction). In a less extent combined plants are used for desulphurization and $NO_x$ reduction of exhaust gases, DESONOX and active carbon plants. Selective Catalytic Reduction occurs at temperatures of 300-400 °C in the presence of catalysts, by injecting ammonia ($NH_3$). Catalysts used: $TiO_2$, $WO_3$, $V_2O_5$.

The rate of reduction of nitrogen oxides is 80-95% for SCR installations. Selective non-catalytic reduction occurs at temperatures of 900-1000 °C in the absence of catalysts, by ammonia injection. The rate of reduction of nitrogen oxides is 30-50%.

The reactions presented in the literature are selective, indicating oxidation of ammonia and sulfur dioxide ($SO_2$) may not occur, the presence of oxygen being essential in the development of some reactions.

Treatment of exhaust gases by passing them in the activated carbon reactor is done at temperatures of 100-150 °C.

Depending on the type of installation, old or (new), by applying BAT (best available techniques), the emission levels of $NO_x$, calculated as $NO_2$, in mg/Nm³, are limited to 100 (100) for boilers, operating on gaseous fuels (calculated at 3% $O_2$), to 90 for gas turbines (calculated at 15% $O_2$) and to 100 for internal combustion engines, stationary, for gaseous fuel, calculated at 15% $O_2$. In case of using liquid fuels on boiler installations, depending on the type of installation, old or (new), and its power by applying the BAT techniques, emission levels of $SO_2$, in mg/Nm³, calculated at 3% $O_2$, are limited to 350 for power below 100MWt, 250 for power below 300MWt and 200 for power over 300MWt.

The amount of $NO_x$ emitted in the means of transport sector in European Union countries (EU-27) represent 46.3% of the total of $NO_x$ emissions at their level (electricity and heat generation sector represent 20.5% and energy using industry represents 13.9%). Due mainly to use low $NO_x$ combustion technology, to the replacement of solid fuels to natural gases, to the use flue gases treatment (selective non-catalytic and selective catalytic) emissions of $SO_2$ decreased from approx. 16900 kt, in 1990, with 31% to about 10300 kt in 2007.

## Solid Particles

Particles released into the atmosphere come both from combustion processes and the entrainment by the combustion gases from the raw material, e.g. rotary drum furnaces for clinker, limestone, dolomite, etc.

Result of the combustion processes, the black smoke is a heterogeneous mixture and variable in structure, between soft solid particulates, water vapors and gases that result from the incomplete combustion of fuels. As a result of the reduced size of suspended solid particles, the smoke refract light and color depending on the concentration and color of substances component in it. That is why the smoke comes in large range of dark colors from the weak gray to black, depending especially by the degree of perfection of the combustion and the quality of burned fuels. The determination of the index that characterizes smoke color is used for technical control of combustion quality, in combating environmental pollution, resulted in particular from burning of solid fuels and lower oils (Bacharach no.). Generally, the formation of unburned solid materials in fossil fuels flames is pointing out especially the wrong mixing between fuel with air that is insufficient to achieve combustion even before carbon oxide is obtained. To liquid fuels without mineral content these solid particles can virtually disappear by burning that if the oxygen is in adequate proportions, in warm enough areas of the furnaces and for a sufficient residence time. In black smoke it is noticed the emergence of two categories of unburned solid: very soft particles of soot and particles much larger than the first, sometimes called cenospheres from intermediate and heavy fuel combustion, or especially flying coke by combustion of solid fuels.

The mechanism of the formation process of soot particles in gaseous fuels flames is explained by theories that belong to three main groups. Thus, the first group includes theories that explain formation by thermal decomposition of hydrocarbons in fuel in carbon and hydrogen, with subsequent polymerization of the carbon. The second group of theories explains soot formation by oxidation of hydrocarbons until peroxides are formed and then by decomposition of peroxides free radicals are divided, which favors higher hydrocarbon formation, which under the high temperatures in the flame, are decomposing to form solid carbon particles. The third group of theories involves the formation of soot particles in flame by polymerization of the $C_2$ radical.

All these theories have a qualitative character, experiments verifying only certain assumptions that are made. The main factors that prompt the formation, quantity, quality and size of soot particles are as following:

- The degree of fuel enrichment (dosage), or lack of required air for stoichiometric combustion;

- Physico-chemical properties of fuels, determined by the gravimetrically report C/H of carbon and hydrogen from fuels and by the molecular mass;

- The conditions of mixing fuel with combustion air which may be sufficient but unevenly distributed in relation to fuel gas;

- The temperature level in the combustion chamber, as the temperature rises the polymerization and hydrocarbon cracking processes intensify;

- The pressure in the combustion chamber which, by increasing, it intensifies the formation of soot process in flame.

The burner of industrial liquid fuels create the fuels spraying as drops, firstly, to increase their surface in contact with the oxidizer. Spraying of liquid fuels in the form of drops jets that ignite in a gaseous environment that lacks the oxygen required for complete combustion, can generate soft soot particles and relatively large solid particles with average diameter up to 100 μm. These last ones having a spongy aspect and spherical form, thus called cenospheres, are more numerous as the liquid fuel is more heavy, meaning that it contains molecules with more carbon atoms.

The cenospheres result from cracking hydrocarbons in liquid phase as soft drops, through decomposition reactions that lead to release of gaseous products and building of solid residue rich in carbon, but which also contains minerals from the original fuel. Similarly, the flying coke is made from soft coal particles, sprayed with air blast at coal dust operating burners. The heavy liquid fuels, in particular, build solid tailings from sediments and organ-metallic compounds that associate with carbon cracking residues. Overall mass of solid particles thus formed for heavy fuels may be in the order of a few decigrams per kilogram of fuel.

Cenospheres are in smaller numbers than particles of soot, but their overall mass is generally much bigger. Essentially, the presence of these relatively large solid particles, sometimes make the regulations compliance on pollution from the combustion of heavy liquid fuels difficult.

Given the size of cenospheres, their combustion is slower than the combustion of soot particles. This combustion is essentially controlled by the diffusion phenomena. Thus, its influence is decisive on: diffusion rate of oxygen to particle, diffusion rate of gases resulted from combustion by the ambient gaseous environment, the particle temperature and the activation energy of substances at work. It should not be missed the fact that the hydrocarbon skeleton of cenospheres contains mineral elements originally present in the fuel. These elements may be particularly harmful in contact with the heating furnaces charge, with the elaboration furnaces bath, refraction bricks from the furnace's walls, boilers or chemical industry ovens pipes. A negative example is the harmful action of vanadium salts on steel at temperatures above 600 $^\circ$C. In flame, the vanadium is present mainly as $V_2O_5$ because $V_2O_4$, $V_2O_3$ and $V_3O_5$ oxides can be considered as intermediate steps to $V_2O_5$. When using a good quality combustion plant and properly exploited, the solid particles mostly disappear by burning, leaving behind ashes of the initial mineral mass of fuel. To the heat generators operating on industrial liquid fuels is imposed that solid suspensions should not exceed 100-200 mg/kg of fuel. This prevents the excessive pollution of the environment and the excessive increase of maintenance costs associated with deposition cleaning on the heat change pipes. Oil use with high sulfur content, requires operation at low excess of air close to the stoichiometric value, to reduce corrosion at low temperature. It is favored this way the increase of losses by unburned particles related to the soot and cenospheres and flying coke respectively. Reducing these polluting emissions is directly influenced by the ratio between carbon C and hydrogen H of the initial fuel and the presence of asfaltenes in oil.

Gaseous fuels with a high ratio C/H produce soot easier than those with the lower C/H ratio. Soot production in the flame makes this to be brighter and to transfer more useable heat by radiation. As a result of this advantage, there was developed a process (called self - carburizing) by which carbon particles produced in the initial zone of the flame are completely oxidized to the top of the flame. This procedure has been used on some burners of boilers and furnaces, proving at the same time with the increasing efficiency very effective to reduce the emissions of CO, $NO_x$ and partially oxidized hydrocarbons.

Primary procedures for reducing particles emissions, are to replace the solid and heavy liquid fuels (oil) with light liquid fuel combustions and to use complete combustion methods.

Primary processes that reduce particle emissions resulting from incomplete combustion of liquid fuel are:

- Prior physical treatment of liquid fuel consisting in preheating, filtration, centrifugation to remove the largest part of sediments, water, soft particles of minerals;

- High quality spraying of which are essential choosing of optimal spraying system, well adjusted preheat temperature of the fuel. After spraying by mechanical injection under pressure of heavy liquid fuels, a variation of the viscosity of dozen centistokes modify essentially the spraying features, with great repercussions on the process of combustion;

- Fuel combustion with preheated air with a moderated air excess (below 30%), using especially overheated steam spraying;

- Improvement, by addition, of the combustion characteristics of industrial fuels used;

- Use of modern combustion techniques such as aerosols combustion, or water - liquid fuels emulsions combustion (with 3% - 5% water).

Secondary procedures for reducing particles emissions, most used by boilers and furnaces are:

- Dry mechanical refinement, by which the particles settling is done under the action of gravity and/or centrifugal force;

- Wet refinement by passing the gases under a liquid mist;

- Refinement by filtration, by passing gases through porous materials;

- Electrical refinement by gas ionization and dispersed particles sedimentation in a high-voltage electrical field. There are considered BAT techniques the use of electrical filters (ESP), textile fabric filters (FF), the wet scrubbers and cyclones, when the latter are used together with ESP or FF. The retention efficiency of ESP for particles larger than 1μm, is 96.5%, the FF is 99.95%, the wet scrubber is 98.5%, and the cyclone is over 85% for particles larger than 5μm.

Depending on the type of installation of boilers, old or (new), also on its power, by applying the BAT techniques, the levels of particles emission, in mg/Nm³, are limited to 30 for powers below 100MWt, 25 for powers below 300MWt and 20 for powers over 300MWt, for liquid fuel (calculated at 3% $O_2$) and for solid fuels (coal and lignite), calculated at 6% $O_2$. For internal combustion engines, stationary, liquid fuel, BAT associated filtering techniques can lead to a maximum value of 30 mg/ Nm³, gas oil and 50 mg / Nm³, heavy liquid fuel.

The amount of PM 10 particles, with diameters less than 10μm, in European Union countries (EU-27), as primary products (directly discharged into the atmosphere) and secondary (derived from the precursors $NO_x$, $SO_2$ and $NH_3$, transformed by photochemical reactions into particles) were reduced by 53% between 1990-2005. The direct amount of PM10 were reduced from 4779 kt in 1990 to 2491 kt in 2005, and the secondary from 33588 kt to 19629 kt. The sector of combustion processes represents 70% of the overall direct emissions of particulates and 90% of the total secondary products. The sector of energy using industry represented 39% of the overall direct emissions of PM10 particulates in 1990 and reached at 13% in 2005, the industry of power generating represented 14% of the overall emissions of PM10 particles in 1990 and reached 10% in 2005, and the means of transport reached from 13% to 21%. The sector of energy using industry represented 39% of the total secondary emission of PM10 particles in 1990 and reached 26% in 2005, the industry of power generation represented 15% of the total secondary emissions of PM 10 particles in 1990 and reached 11% in 2005, and the means of transport reached from 26% to 29%. These substantial reductions were done, due to the dusting techniques of the exhaust gases, the replacement of solid and liquid fuels with high content in sulfur with natural gases, the introduction of the primary techniques for achieving combustion with reduced $NO_x$ emissions, the use of flue gases treatment and the application in EU-27 of new regulations which limited the content of sulfur from some liquid fuels and which used advanced catalysts in internal combustion engines for transportation.

## Volatile Organic Compounds (VOCs) and Persistent Organic Pollutants (POPs)

Volatile organic compounds are aliphatic and aromatic hydrocarbons, alcohols, ketones, esters, aldehydes, benzene, toluene, acetones, methanol and formaldehydes.

POPs are mainly composed of polycyclic aromatic hydrocarbons (PAHs), resulted in combustion processes of fuels in a smaller extent, but with particularly harmful effects, from dioxins and furans, resulted from the incineration of waste, pesticides and products of chemical industry.

Partially oxidized hydrocarbons are generated by the lack of oxygen (combustion air) in combustion process, or from slow oxidation.

Lack of oxygen in the combustion process can be explained locally or temporarily, because combustion installation from industry are supplied with excess of combustion air. Due to poor mixing of fuel-combustion air, in the flame can occur rich zones in fuels which are formed unsaturated hydrocarbons, aldehydes and acids and, in some cases, the process can continue until the formation of soot. These unburned elements may evolve, behind the flame front, depending on thermodynamic parameters and the interaction of oxidants.

To achieve sufficiently high oxidation rates the following are necessary:

- To set the mixture of fuel gas - combustion air between flammability limits;

- Achieving a level of temperature high enough ahead of the flame front;

- Creating a bigger turbulence;

- Ensuring that the various intermediate elements stay a sufficient period of time in the flame.

In combustion processes, the objective of reducing VOCs is their volatilization and oxidation to $CO_2$ and $H_2O$, by thermal oxidation, catalytic or bio-oxidation. In the case of incinerators, especially, is practiced burning in secondary outbreaks, in which the catalysts can be introduced, after the primary focus.

Among the sectors of fuel combustion processes that produce VOCs, transport is the most significant with 18,6% of total EU emissions, the rest being insignificant. Between 1990 and 2007, road transport has reduced VOCs by 60%, as a result of using the three-way catalytic converter and the improvement of diesel quality.

Among the sectors of fuel combustion processes that produce PAHs, residential heating sector is the most significant with 41% of total EU emissions, and transport represents 15%. Between 1990 and 2007, residential heating sector reduced the production of PAHs by 14% and road transport by 47%, mainly due to the advanced catalyst.

## Atmospheric Greenhouse Gas Concentrations

Among the gases that produce greenhouse effect, the carbon dioxide has the most important contribution, its percentage being 65%. It appears in an overwhelmingly proportion after fuel combustion processes. In the year 2009 the concentration of $CO_2$ in the atmosphere was 387ppm. Carbon dioxide recently considered as pollutant, prevents the discharge of heat energy from the surface of the Earth which is acquired by radiation from the sun. As the time passes, even the increase with a few degrees of the temperature could lead to reduction of the land ice cap, the spread of deserts, raising sea and oceans levels, as well as important climate changes. The concentration of $CO_2$ in the atmosphere has increased by 38% compared to the pre-industrial age, when it was 280 ppm.

In 1990, a report of a United Nations committee has confirmed the tendency of heating of the globe and considered that, if the world will not take measures to stabilize the gas flow released in the atmosphere (carbon dioxide, methanol, gases from chemical fertilizers, various hydrocarbons), the temperature will rise by three degrees until 2000. A United Nations Framework Convention on Climate Changes was negotiated between 1991 - 1992 and signed at Rio Conference in 1992. The Convention entered into force in 1995. The main objective of the Convention was to stabilize the greenhouse gas concentrations from atmosphere, especially carbon dioxide, at a level that will not endanger climate system.

Kyoto Protocol, signed in 1997, with the exception of USA, is an extension of the UNFCCC (United Nations Framework Convention on Climate Change) from 1993 and is the first step in an effort to reduce the long-term changes in emissions to prevent climate changes. Both at the global and the EU level this "dangerous anthropogenic interference" has been recognized by formulating the objective of keeping the long-term global average temperature rise below 2 °C compared to pre-industrial times.

There is a consensus in EU-27 countries for collective action to reduce greenhouse effect gas emanations by: improving techniques for burning fossil fuels, the use of renewable sources for energy production (water, sun, wind, geothermal energy), improving industrial technologies, decreasing the consumption of fuels for road transport, increasing the share of public transport, reduce emanation of animal manure, reduce the fertilizers based on nitrogen, stopping the clearance of forest and accelerated action of forestation, reducing waste and gases from municipal waste dumps, etc. The commitment taken by the EU-27 for 2008-2012 compared to 1997, is a reduction of 8% versus 5%, as was stipulated in the Kyoto Protocol.

## Determination of Pollutant Emissions in Combustion Processes

Quantitative determination of pollutant emissions refers to:

- The determination of pollutant concentration, $c_i$ [mg/Nm³], representing the concentration of pollutant i in the gases released in atmosphere, required to screening verification in the emission limits stipulated by the regulations;

- Determination of pollutant quantity emission, which represents the amount of pollutants released into the atmosphere over a specified period, usually one year, $E_i$ [t], required to establish the activities contribution from combustion processes at general air pollution.

Determination of pollutant concentration, $c_i$ [mg/Nm³] and the amount of pollutants $E_i$ [t] produced by a combustion plants requires knowledge of the effluent gaseous composition in point of measurement. In addition, at determination of pollutant quantity emission, it is necessary to know the amount of fuel consumed during the period, in m³N, for gaseous fuels, or kg for liquid or solid fuels.

The measurements are based on national norms, which are prescribed the measuring and calculation methodologies, type of measurement devices that are accepted, as well as the execution range for measurements. For large combustion plants requires continuous monitoring. The automatic devices that are used:

- Gas analyzer to determine the volumetric concentrations, from the flue gas mixture, $O_2$ [% volumetric], $CO_2$ [% volumetric], CO [ppm], $NO_x$ [ppm], $NO_2$ [ppm], and $SO_2$ [ppm];

- Automatic device to determine the volatile organic compounds;

- Automatic instrument, with isokinetic probe for determining particulate matter.

Gaseous effluent concentrations measured values are required, on the one hand to determine the quantity of emissions of pollutants and on the other hand directly to determine the concentration (in mg/Nm³) of the emission of pollutants. Calculation of concentrations of different pollutants emissions i, $c_i$[mg/Nm³], related to reference conditions, it is required to be done to compare it with the limit values set by national regulations, for periods of time, depending on the type, age and power plant and fuel type consumed.

E.g. if the analyzer indicates a concentration of 60 ppm $SO_2$, 80 ppm $NO_x$ and 10.5% $O_2$, to measurement of gaseous effluents produced by burning a liquid fuel, it can be calculated the concentration of pollutant emission of $SO_2$, $CSO_2$, $NO_x$, $C_{Nox}$, respectively.

$$c_{SO_x}\left(mg/Nm^3\right) = c_{SO_x mas}\left(ppm\right) \cdot k1 \cdot \frac{21 - c_{O_2 ref}}{21 - c_{O_2 mas}}$$

$$c_{NO_x}\left(mg/Nm^3\right) = c_{NO_x mas}\left(ppm\right) \cdot k2 \cdot \frac{21 - c_{O_2 ref}}{21 - c_{O_2 mas}}$$

where:

$CSO_{2\ mas}$ - $SO_2$ emission concentration measured in ppm;

$CNO_{x\ mas}$ - $NO_x$ emission concentration measured in ppm;

k1 – 2.85 – ppm conversion factor mg/Nm³ for $SO_2$;

k2 – 2.05 – ppm conversion factor mg/Nm³ for $NO_2$;

$CO_{2\ ref}$ – 3% volumetric - reference oxygen concentration for a gaseous fuel according to EU norms;

$CO_{2\ mas}$ – measured oxygen concentration in the gaseous effluent, in volumetric %.

Replacing in $c_{SO_x}\left(mg/Nm^3\right) = c_{SO_x mas}\left(ppm\right) \cdot k1 \cdot \frac{21 - c_{O_2 ref}}{21 - c_{O_2 mas}}$ with numeric values it is obtained:

$$c_{SO_x} = 60 \cdot 2,85 \cdot \frac{21 - 3}{21 - 10,5} = 293,1\, mg/Nm^3$$

$$c_{NO_x} = 80 \cdot 2,05 \cdot \frac{21 - 3}{21 - 10,5} = 281,1\, mg/Nm^3$$

The amount thus calculated is compared with the emission limit for $SO_2$ and $NO_x$ respectively, according to national norms, which are usually set for some period of time, depending on the type, age and power of the plant and fuel type consumed.

In the case of large combustion plants, which consume liquid fuels, in table are presented associated BAT limit values, for $NO_x$ emissions calculated as $NO_2$ and $SO_2$, in mg/Nm³, for 3% $O_2$, depending on the type and power of the plant.

Table: Limit values of $SO_2$ and $NO_x$ emissions for electricity and heating generation.

| Crt. No. | Plant type | Pollutant | Measure unit | Thermal power, MWt | | |
|:---:|:---:|:---:|:---:|:---:|:---:|:---:|
| | | | | <100 | 100-300 | "/>300 |
| 1 | Old | $SO_2$ | mg/Nm³ | 350 | 250 | 200 |
| 2 | Old | $NO_x$ | mg/Nm³ | 450 | 200 | 150 |
| 3 | New | $SO_2$ | mg/Nm³ | 350 | 200 | 150 |
| 4 | New | $NO_x$ | mg/Nm³ | 300 | 150 | 100 |

It is found that the values calculated for the exemplified installation would not exceed the limits unless it would have a thermal power less than 100 MW.

To determine the exact amount of a pollutant i, Ei, in t for a certain period, the general relation can be used:

$$E_i = ki \cdot V_{gu} \cdot B \cdot c_i \cdot 10^9 \, [t]$$

where:

$ki$ – conversion factor for the i pollutant, ppm mg/Nm³.

$V_{gu}$ – volume of dry burned gases produced by burning the fuel, in m³N/m³N or m³N/kg.

$B$ – fuel consumption, in Nm³ of consumed gas in the respective period, or kg of liquid or solid fuel.

$c_i$ – measured concentration of i pollutant, in ppm.

The specialty papers shows calculating relations for the volume of dry flue gases resulting from combustion of various fuels. For example, for combustion of gaseous fuels to find the volume of dry flue gases produced from burning a Nm³ of fuel, use the following formula:

$$V_{gu} = \frac{CO_2^{(c)} + CO^{(c)} + \sum mC_m H_n^{(c)}}{CO_2^{(g.a)} + CO^{(g.a)} + \sum mC_m H_n^{(g.a)}}$$

Values in the numerator results from the initial analysis of fuel gas composition and the denominator values are measured at the exhaust of plant flue gases, for example with a flue gas analyzer.

For less accurate calculations, but faster, it is used the relationship:

$$E_i = fi \cdot B \cdot PCI \cdot 10^6 \, [t]$$

where:

$fi$ – emission factor, in g/GJ;

$B$ – fuel consumption, in Nm³ of consumed gas in the in the respective period, or kg of liquid or solid fuel;

$PCI$ – low calorific power of the fuel, in GJ/Nm³, GJ/kg respectively.

The factors $fi$ (in g/GJ, respectively lb/$10^6$ Btu), are used in the EU and U.S. respectively, to calculate the amount of pollutants emitted into the atmosphere over a period of time and are tabled according to the many types of fuels, installation, etc. In Table are summarized the approximate amounts of various pollutants to be emitted into the atmosphere worldwide, calculated with the emission factors $fi$ (taking the values for coals, oils and natural gas) for fuel consumption in 2030 provided by the World Energy Outlook -2008.

Table: Approximate amounts of various pollutants to be emitted into the atmosphere worldwide, calculated with emission factors fi, for fuel consumption in 2030.

| crt. no. | Fuel type | | Pollutant | | | | |
|---|---|---|---|---|---|---|---|
| | | | $SO_2$ | $NO_x$ | CO | $PM_{10}$ | $VOC_S$ |
| 1 | Coals | Emission factor [g/GJ] | 10746.5 | 4083.67 | 1308.49 | 3094.99 | 21.49 |
| | | Pollutant quantity [kt] | 2207331 | 838786 | 268766 | 635713 | 4414 |
| 2 | Oils | Emission factor [g/GJ] | 7307.62 | 1590.48 | 143.57 | 343.89 | 38.68 |
| | | Pollutant quantity [kt] | 1562457 | 340064 | 30697 | 73528 | 8270 |
| 3 | Natural gas | Emission factor [g/GJ] | 0.0 | 601.80 | 150.88 | 12.89 | 25.79 |
| | | Pollutant quantity [kt] | 0.0 | 92430 | 23173 | 1979 | 3961 |

To these annual amounts of pollutants emission the amount of $CO_2$ is added, approximate for the year 2030, to approx. 30 million kt, which would lead to a planetary temperature exceeding by more than 2 °C. To reduce the quantity of air pollutants emissions it is necessary that reduction measures and techniques to be generalized on a planetary scale Also there should be mentioned the EU-27 achievements in the period 1990-2007, that have significantly reduced emissions of pollutants into the atmosphere: CO by 63%, $NO_x$ by 31%, $SO_2$ by 69%, VOCs by 41%, PAHs by 63% and PM10 by 53%. For the future, the key to success in reducing the air pollution is increasing energy efficiency and using regenerative resources. In 2009, the EU agrees a Climate and Energy Package to:

- Reduce greenhouse gas emission by 20% by 2020;

- Increase the share of renewable energy by 20% by 2020;

- Improve energy efficiency by 20% by 2020.

## References

- Air-pollution-control-devices, encyclopedia: energyeducation.ca, Retrieved 13 Febuary, 2019

- "Baghouse / Fabric Filters KnowledgeBase". Neundorfer.com. Archived from the original on 2013-08-07. Retrieved 2013-09-08

- IUPAC, Compendium of Chemical Terminology, 2nd ed. (the "Gold Book") (1997). Online corrected version: (2006–) "electrostatic precipitator". doi:10.1351/goldbook.E02028

- Farnoud A (2008). Electrostatic Removal of Diesel Particulate Matter. ProQuest. p. 23. ISBN 978-0549508168

- Scrubber, encyclopedia: energyeducation.ca, Retrieved 4 January, 2019

- Pardon G, Ladhani L, Sandstrom N, et al. (2015). "Aerosol sampling using an electrostatic precipitator integrated with a microfluidic interface". Sensors and Actuators. B, Chemical. 212: 344–352. doi:10.1016/j.snb.2015.02.008

- Gasscrubber, gas-purification-techniques, air-purification: lenntech.com, Retrieved 15 June, 2019

- "Thermal Oxidizer". U.S. EPA Technology Transfer Network Clearinghouse for Inventories & Emissions Factors. U.S. Environmental Protection Agency. Retrieved 4 April 2015

- Rasmussen, Søren (2006). "Characterization and regeneration of Pt-catalysts deactivated in municipal waste flue gas". Applied Catalysis B: Environmental. 69: 10–16. doi:10.1016/j.apcatb.2006.05.009

- Reduction-of-air-pollution-by-combustion-processes, the-impact-of-air-pollution-on-health-economy-environment and agricultural-sources, books: intechopen.com, Retrieved 4 Feb, 2019

# Index

## O
Ozone Depletion, 1, 90, 115, 121-123, 125-133
Ozone Hole, 24, 121-122, 124-128, 131, 133-134
Ozone Injury, 147-149

## P
Perfluorocarbons, 7, 20-21
Peroxyacetyl Nitrate, 2
Polar Stratospheric Clouds, 125-127, 134
Pollution Standards Index, 12

## R
Radiative Forcing, 27, 42, 44-47, 132
Radon Gas, 4, 59-60, 86
Roadway Air Dispersion Modeling, 9, 80-81

## S
Solar Radiation, 26, 42, 44, 47, 85, 100, 126, 132, 213, 217
Stratosphere, 23, 25, 27, 47, 122-128, 131-134
Stratospheric Ozone, 1, 27, 90, 126, 128-129
Sulfate Aerosol, 46
Sulfur Dioxide, 1-2, 6, 12, 14-15, 19, 21, 42, 65, 104-105, 112, 115-116, 118, 120, 138-140, 143-145, 149, 151, 169, 199, 212-214, 223

## Sulfur Hexafluoride, 20-21
Sulfur Hexafluoride, 20-21
Sulfuric Acid, 42, 109, 116, 139-141, 189, 191, 195-197, 212-214
Suspended Particulate Matter, 41

## T
Temperature Sensor, 162-164
Triethanolamine, 155
Troposphere, 23-25, 27-28, 78, 122-123, 132, 134, 159

## U
Urban Airshed Model, 40
Uv Spectrum, 27-28

## V
Volatile Organic Compounds, 2, 4, 13-14, 24, 28, 54, 63, 74, 86, 98, 116, 118, 205, 226, 229
Volcanic Eruption, 2, 103, 112-114

## W
Water Vapour, 46, 68, 146, 199
Wet Scrubbers, 8, 43-44, 174, 180, 187, 200, 202-203, 226
Wind Velocity, 9, 65

www.ingramcontent.com/pod-product-compliance
Lightning Source LLC
Chambersburg PA
CBHW061251190326
41458CB00011B/3646